SOLUTIONS TO PROBLEMS AND STUDY AIDS FOR
ORGANIC CHEMISTRY
A BACKGROUND FOR THE LIFE SCIENCES
GARDNER W. STACY
Washington State University

Harper & Row, Publishers
New York   Evanston   San Francisco   London

Solutions to Problems and Study Aids for ORGANIC CHEMISTRY: A Background for the Life Sciences

Copyright © 1975 by Gardner W. Stacy

All rights reserved. Printed in the United States of America. No part of this book may be used or reproduced in any manner whatsoever without written permission except in the case of brief quotations embodied in critical articles and reviews. For information address Harper & Row, Publishers, Inc., 10 East 53rd Street, New York, N.Y. 10022.

Standard Book Number: 06-046389-9

CONTENTS

A Review of Organic Reactions   1

Answers to Problems   35

  1. Background and Introductory Concepts   36

  2. Valence and Fundamentals of Structure   38

  3. Classes of Organic Compounds and Nomenclature   45

  4. Alkanes—Reactive Carbon Intermediates   55

  5. Alkenes and Alkynes—Addition Polymerization   60

  6. Stereochemistry—Geometric and Optical Isomerism   68

  7. Alicyclic Hydrocarbons and Derivatives—The Steroids   76

  8. Arenes—Benzene-Type Hydrocarbons   82

  9. Organic Halides and Nucleophilic Substitution—Organometallics   89

10. Alcohols, Phenols, Ethers, and Related Sulfur Compounds   94

11. Amines—Dyes—Alkaloids   100

12. Aldehydes and Ketones   109

13. Carbohydrates—Sugars and Polysaccharides   118

14. Nucleic Acids—Structure and Molecular Biology   128

15. Carboxylic Acids and Related Substances   133

16. Glycerides—Fats and Oils   144

17. Peptides and Proteins (Amino Acid Polymers) 151

18. Natural Products and Physiologically Active Substances 161

19. Determination of Organic Structures from Spectral Properties 169

A REVIEW OF ORGANIC REACTIONS

This section summarizes the general features of all the reactions and preparation of the various organic families described in the text. An effective way to review these reactions is to write them out using specific alkyl and aryl groups. It will often be helpful to start with the simplest examples in which R stands for CH_3 or C_6H_5, where appropriate. Then, write out more examples using different alkyl, alicyclic (Chapter 7), and aryl (Chapter 8) groups. The section or reaction number indicating where each reaction is discussed in the text is indicated for easy reference.

PREPARATION OF ALKANES

1. Reduction of alkenes (4.3.1)

    $$RCH=CHR \xrightarrow[\text{heat, pressure}]{H_2,\ Ni} RCH_2CH_2R$$

    Same carbon skeleton as in starting material.

2. Reduction of halides (4.3.2)

    $$R-X \xrightarrow[Zn,\ HCl-CH_3CO_2H]{[H]} R-H\ +\ HX$$

    Same carbon skeleton as in starting material.

3. Alkene coupling (4.3.3)

    $$2\ RCH=CH_2 \xrightarrow[\text{2. KOH, AgNO}_3]{\text{1. }B_2H_6} RCH_2CH_2-CH_2CH_2R$$

    Carbon skeleton in product twice that of starting material.

REACTIONS OF ALKANES

1. Oxidation (4.5.1)

$$RH \xrightarrow[\text{heat}]{O_2} CO_2 + H_2O$$

Carbon skeleton completely fragmented.

2. Nitration (4.5.2)

$$RCH_2R \xrightarrow[\text{heat}]{HONO_2} RCH_2NO_2 + RNO_2$$

Carbon skeleton fragments to various sized nitroalkanes.

3. Halogenation (4.5.3)

$$RH + X_2 \longrightarrow RX + HX$$

$$X = Cl, Br$$

Hydrogen is substituted by halogen— a substitution reaction. The reverse of Reaction 4.3.2.

4. Cracking or pyrolysis (4.13, 4.14)

$$RCH_2CH_2CH_2CH_3 \xrightarrow[500°]{Al_2O_3\text{-}SiO_2} RCH_3 + CH_2=CHCH_3$$

Carbon-carbon bond broken to give shorter alkane and propylene, butylene, or isobutylene. Carbon-hydrogen bonds can also be broken.

5. Isomerization (4.15)

$$RCH_2CH_2CH_3 \xrightarrow[\text{heat}]{AlCl_3} RCHCH_3 \atop \ \ \ \ \ \ \ \ \ \ \ \ \ \ |\ \atop \ \ \ \ \ \ \ \ \ \ \ \ \ CH_3$$

Rearrangement of the C skeleton.

6. Alkylation (4.16)

$$(CH_3)_3CH + H_2C=CHR \xrightarrow{H_2SO_4} (CH_3)_3CCH_2CH_2R$$

Addition of an alkane to an alkane.

7. Aromatization (4.17)

$$CH_3(CH_2)_5CH_3 \xrightarrow[\text{heat}]{Pt\ cat.}$$

(toluene: benzene ring with CH$_3$ substituent)

Important industrial reaction for ring formation. Although aromatic nuclei are formed, the reaction cannot be predicted in a general manner for longer chain alkanes.

PREPARATION OF ALKENES AND ALKYNES

FORMATION OF C=C

1. Dehydration of alcohols (5.3.1)

   $-\underset{H\ \ OH}{\overset{|\ \ |}{C-C}}- \xrightarrow[\substack{\text{or} \\ H_2CO_4 \\ \text{heat}}]{Al_2O_3} -\overset{|}{C}=\overset{|}{C}- + H_2O$

   Elimination of elements of water from adjacent C atoms.

2. Dehydrohalogenation of alkyl halides (5.3.2)

   $-\underset{H\ \ X}{\overset{|\ \ |}{C-C}}- \xrightarrow[\text{ethanol as solvent}]{KOH} -\overset{|}{C}=\overset{|}{C}- + HX$

   Elimination of elements of HX from adjacent C atoms.

FORMATION OF C≡C

1. Substitution involving acetylides and alkyl halides (5.5.1)

   $R'C{\equiv}CNa + X-R \longrightarrow R'C{\equiv}CR + NaX$ Substitution reaction.

2. Dehydrohalogenation of vicinal dihalides (Pb. 5.10)

   $R\underset{X}{\overset{|}{C}}H-\underset{X}{\overset{|}{C}}HR' \xrightarrow{NaNH_2} RC{\equiv}CR' + 2HX$

   A double elimination reaction from the same C atoms.

PREPARATION OF DIENES C=C—C=C

1. Double elimination (5.26)

   $XCH_2CH_2CH_2CH_2X \xrightarrow[\text{ethanol}]{KOH} CH_2{=}CH{-}CH{=}CH_2$

   A double elimination reaction from adjacent C atoms.

REACTIONS OF ALKENES (C=C)

1. Catalytic hydrogenation (4.3.1)

   $-\overset{|}{C}{=}\overset{|}{C}- \xrightarrow[\text{also Pd or Pt}]{H_2,\ Ni} -\underset{H}{\overset{|}{C}}-\underset{H}{\overset{|}{C}}-$

   Addition of hydrogen to produce saturated carbon.

3

2. Addition of halogen (5.4.2)

$$-\underset{|}{\overset{|}{C}}=\underset{|}{\overset{|}{C}}- \xrightarrow{X_2} -\underset{|}{\overset{|}{C}}-\underset{|}{\overset{|}{C}}- \\ X \; X$$

$X_2 = Cl_2, Br_2$

Formation of vicinal dihalides.

3. Addition of acids (5.4.3)

$$\underset{R_1}{\overset{H}{\underset{|}{C}}}=\underset{R_3}{\overset{R_2}{\underset{|}{C}}} \xrightarrow{HX} R_1-\underset{\overset{|}{H}}{\overset{\overset{|}{H}}{C}}-\underset{\overset{|}{X}}{\overset{\overset{|}{R_2}}{C}}-R_3$$

HX = HCl, HBr, HI
Addition leads to the formation of alkyl halides. H adds to least substituted C in accord with Markownikoff's rule.

4. Oxidation of alkenes (5.4.4)

$$-\underset{|}{\overset{|}{C}}=\underset{|}{\overset{|}{C}}- + HOH + [O] \xrightarrow{KMnO_4} -\underset{\overset{|}{OH}}{\overset{|}{C}}-\underset{\overset{|}{OH}}{\overset{|}{C}}-$$

Hydroxylation to form 1,2-diols.

5. Coupling of alkenes by diborane (4.3.3)

$$2 \; -\underset{|}{\overset{|}{C}}=\underset{|}{\overset{|}{C}}- \xrightarrow[\text{KOH, then } AgNO_3]{B_2H_6} -\underset{H}{\overset{|}{C}}-\underset{|}{\overset{|}{C}}-\underset{|}{\overset{|}{C}}-\underset{H}{\overset{|}{C}}-$$

REACTIONS OF ALKYNES (C≡C)

1. Addition of hydrogen (5.2)

$$-C \equiv C- \xrightarrow[Pt]{H_2} -CH=CH- \xrightarrow[Pt]{H_2} -CH_2-CH_2-$$

Preparation of alkene intermediate in good yield possible if desired.

2. Reaction with heavy-metal cations (5.6.2)

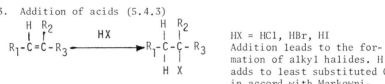

H on a triple bonded C is acidic.

3. Addition of acids (5.6.3)

$$-C \equiv C- \xrightarrow[HgCl_2]{HCl} -\underset{\overset{|}{H}}{\overset{|}{C}}=\underset{\overset{|}{Cl}}{\overset{|}{C}}- \xrightarrow[HgCl_2]{HCl} -\underset{\overset{|}{H}}{\overset{\overset{|}{H}}{C}}-\underset{\overset{|}{Cl}}{\overset{\overset{|}{Cl}}{C}}-$$

Addition occurs in accord with Markownikoff's rule; intermediate alkene can be isolated if desired.

REACTIONS OF CONJUGATED DIENES ($-C=C-C=C-$)

1. Conjugate addition (5.27)

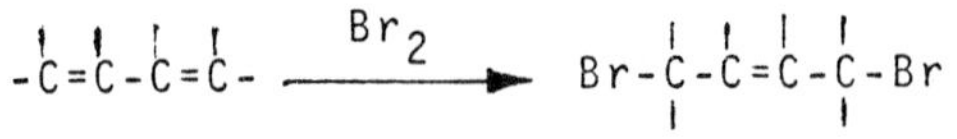

$$-\overset{|}{C}=\overset{|}{C}-\overset{|}{C}=\overset{|}{C}- \xrightarrow{Br_2} Br-\overset{|}{C}-\overset{|}{C}=\overset{|}{C}-\overset{|}{C}-Br$$

Addition to the ends of the conjugated system; the most common type is 1,4-addition.

2. Diels-Alder reaction (5.28)

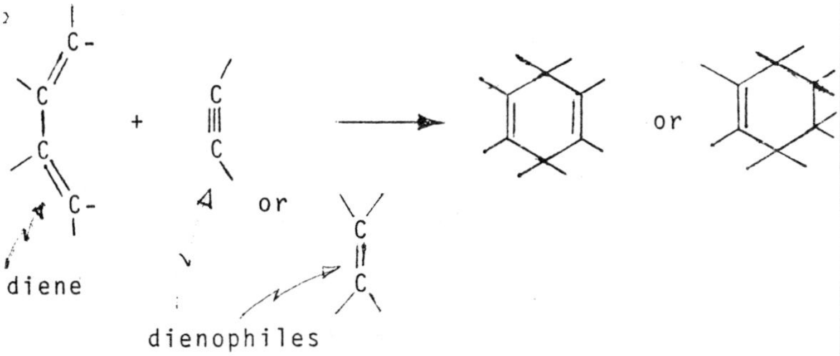

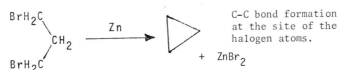

diene   dienophiles

Many dienophiles are made more reactive by attachment of one or more electron-attracting groups.

PREPARATION OF ALICYCLIC COMPOUNDS

1. C_3, Cyclopropanes

   a. From dihalides (7.10)

   $$\begin{array}{c}BrH_2C \\ \diagdown \\ CH_2 \\ \diagup \\ BrH_2C\end{array} \xrightarrow{Zn} \triangle \;+\; ZnBr_2$$

   C-C bond formation at the site of the halogen atoms.

   b. By the Simmons-Smith reaction (7.11)

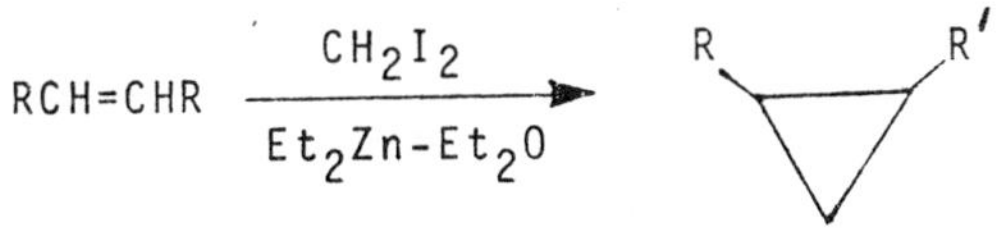

   $$RCH=CHR \xrightarrow[Et_2Zn-Et_2O]{CH_2I_2} \overset{RR'}{\triangle}$$

   Insertion of CH_2 at the double bond.

2. By hydrogenation

   a. C_5, Cyclopentane derivatives from cyclopentadiene (7.8)

   Stepwise reduction of two double bonds from a commercially available starting material.

   b. C_6, Cyclohexanes from benzenes (7.7)

   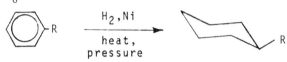

   Complete reduction of the benzene ring.

   c. C_8, Cyclooctane derivatives from cyclooctatetraene (7.14)

   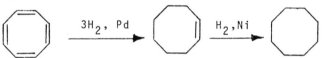

   Stepwise reduction of four double bonds to one, followed by complete reduction to give a saturated ring.

3. C_8, Tetramerization of acetylene (7.14)

   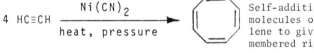

   Self-addition of four molecules of acetylene to give an eight-membered ring.

SOME REACTIONS OF ALICYCLICS

1. Formation of geometric isomers (7.9)

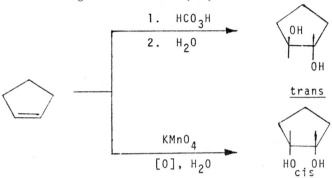

Formation of trans and cis isomers by stereoselective reactions.

2. Reactions of cyclopropane (7.12)

   a. Reaction with hydrogen

   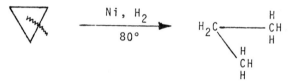

   Opening of the cyclopropane ring with addition of H.

   b. Reaction with bromine

   Opening of the cyclopropane ring with addition of Br.

PREPARATION OF SUBSTITUTED ARENES

1. Friedel-Crafts reaction (Section 8.2)

   a. Alkylation (8.6)

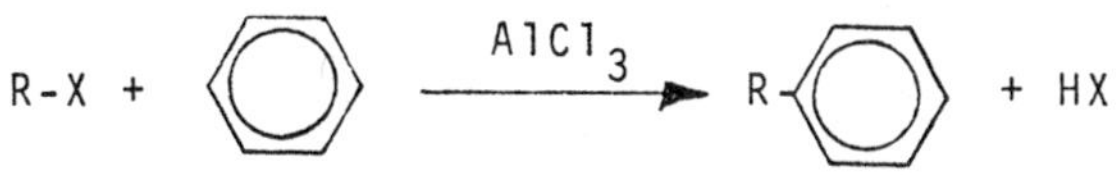

   R-X + ⌬ →(AlCl$_3$) R-⌬ + HX

Reaction of an alkyl halide or alkene with an aromatic hydrocarbon. The rearrangement of a straight chain to a branched chain is a problem in synthesis.

b. Acylation (8.10)

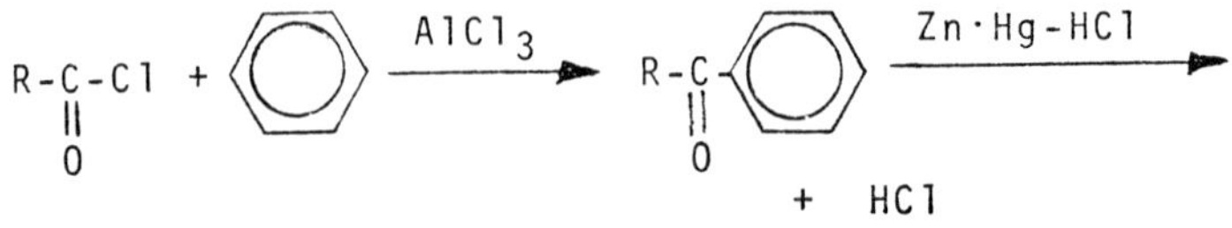

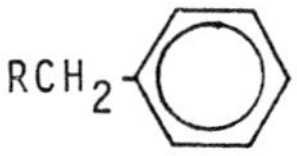

$$R-C(=O)-Cl + C_6H_6 \xrightarrow{AlCl_3} R-C(=O)-C_6H_5 + HCl \xrightarrow{Zn \cdot Hg-HCl}$$

$$RCH_2-C_6H_5$$

Combined with the second step, a Clemensen reduction, this method affords an unrearranged side chain.

REACTIONS OF ARENES

1. Saturation of the ring (8.2)

    a. Hydrogenation

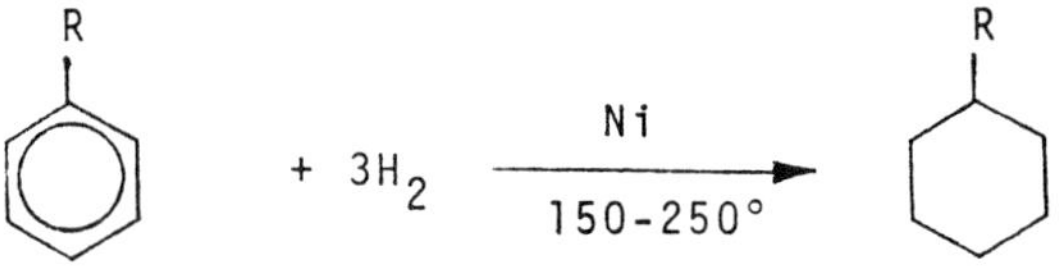

$$R-C_6H_5 + 3H_2 \xrightarrow[150-250°]{Ni} R-C_6H_{11}$$

Useful in preparing cyclohexane derivatives.

   b. Chlorination

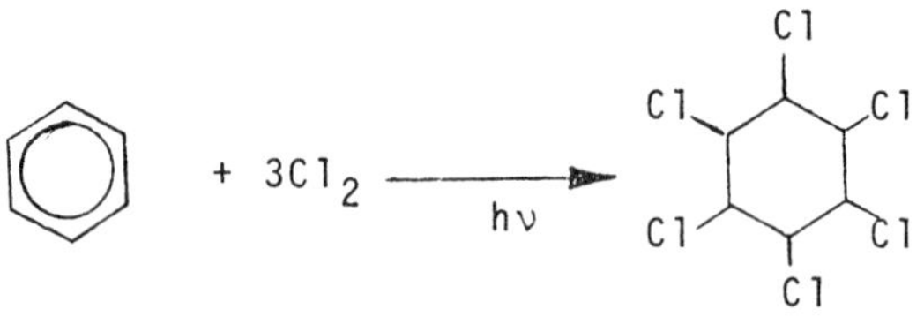

$$C_6H_6 + 3Cl_2 \xrightarrow{h\nu} C_6H_6Cl_6$$

Chlorination and hydrogenation both require more energy to take place than with an alkene.

2. Side-chain oxidation (8.12)

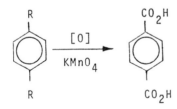

Any number of side chains of varying chain length are oxidized down to carboxyl groups attached to the ring.

3. Electrophilic substitution reactions

   a. Halogenation (Section 8.3.2)

   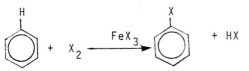

   A reaction useful in introducing one or more halogen groups into arenes or substituted arenes.

   X = Cl, Br

   b. Nitration (Section 8.3.3)

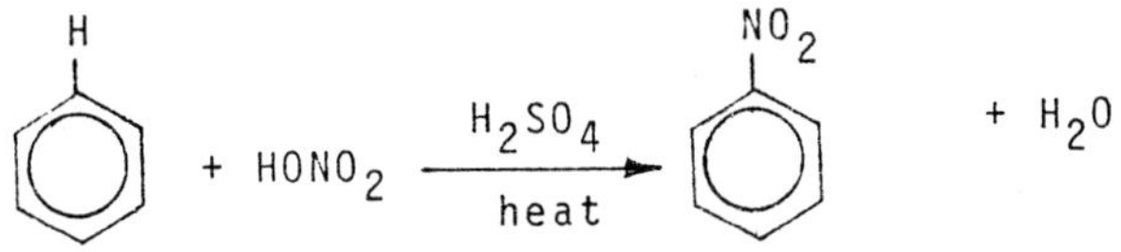

   $$\text{C}_6\text{H}_6 + \text{HONO}_2 \xrightarrow[\text{heat}]{\text{H}_2\text{SO}_4} \text{C}_6\text{H}_5\text{NO}_2 + \text{H}_2\text{O}$$

   The introduction of each successive nitro group becomes more difficult because the NO_2 group deactivates the ring.

   c. Sulfonation (Section 8.3.4)

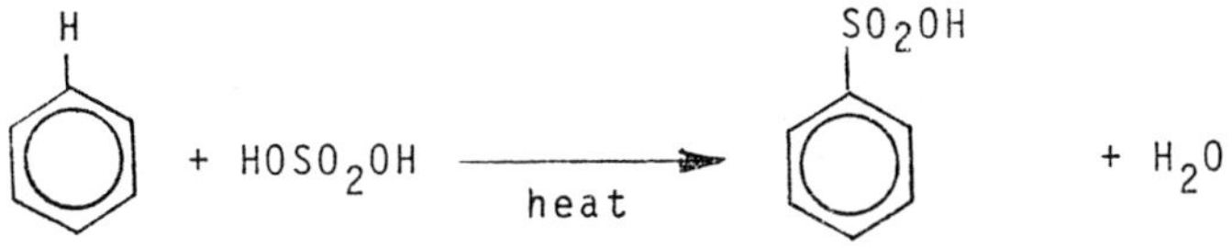

   $$\text{C}_6\text{H}_6 + \text{HOSO}_2\text{OH} \xrightarrow{\text{heat}} \text{C}_6\text{H}_5\text{SO}_2\text{OH} + \text{H}_2\text{O}$$

   The most important application of this reaction is in the manufacture of synthetic detergents.

   d. Friedel-Crafts reaction (Section 8.2)
      See the Review Box above on the preparation of substituted arenes.

PREPARATION OF ORGANIC HALIDES

1. Reaction of an alkane or cycloalkane with X_2, X = Cl, Br (4.5.3)

   $$RH \xrightarrow{X_2} RX + HX$$

   Only symmetrical hydrocarbons with all equivalent carbons will give a single product in this substitution reaction.

2. Reaction of side-chain arenes with X_2 (8.3.2, Pb. 9.2)

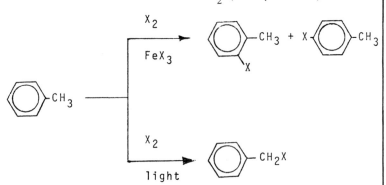

A difference in reaction conditions can sometimes lead to entirely different products, in this case substitution of the ring as compared to the side-chain.

3. Addition of halogen to alkenes and alkynes (5.4.2., Pb. 5.12)

$$RCH=CHR' \xrightarrow{X_2} \underset{\underset{X\ \ X}{|\ \ |}}{RCH-CHR'}$$

Addition of halogen to a double bond gives 1,2-dihalo compounds.

$$RC \equiv CR' \xrightarrow{X_2} \underset{\underset{X\ \ X}{|\ \ |}}{RC=CR'} \xrightarrow{X_2} \underset{\underset{X\ \ X}{|\ \ |}}{\overset{\overset{X\ \ X}{|\ \ |}}{RC-CR'}}$$

Stepwise addition of halogen to a triple bond gives a 1,2-dihalo alkene which can be isolated. The final product is a 1,1,2,2-tetrahalo compound.

4. Addition of hydrohalogen acid to alkenes and alkynes (5.4.3, Pb. 5.12)

$$RCH=\underset{\underset{}{|}}{\overset{\overset{R'}{|}}{C}}-R'' \xrightarrow{HX} RCH_2-\underset{\underset{X}{|}}{\overset{\overset{R'}{|}}{C}}-R''$$

Hydrohalogen acid adds in accord with Markownikoff's rule.

$X = Cl, Br, I$

$$RC \equiv CR' \xrightarrow{HX} \underset{\underset{X}{|}}{RCH=CR'} \xrightarrow{HX} RCH_2-\underset{\underset{X}{|}}{\overset{\overset{X}{|}}{C}}R'$$

In adding HX to an alkyne, one can isolate the intermediate haloalkene.

5. Reaction of alcohols with hydrohalogen acid (9.1.2)

$$ROH \xrightarrow{HX} RX + H_2O$$

$X = Cl, Br, I$

Relative reactivity of acids: $HI > HBr > HCl$; relative reactivity of alcohols: $3° > 2° > 1°$, $ZnCl_2$ catalyst needed with 1° alcohol and HCl.

6. Introduction of fluorine

   a. Perfluorination (Reaction 9.3)

   $$CH_3(CH_2)_xCH_3 \xrightarrow{CoF_3} CF_3(CF_2)_xCF_3$$

   Complete substitution of all H atoms in the hydrocarbon by F.

   b. Substitution of X by F (Reaction 9.4, Pb. 9.9, Reactions 9.5 and 9.6)

   $$RX \xrightarrow[\text{or } HgF_2]{KF} RF$$

   These substitution reactions introduce F in place of another halogen atom.

   $$CHCl_3 \xrightarrow{HF-SbF_5} CHClF_2$$

   c. Addition of HF and [2F] to alkenes (Pbs. 9.7, 9.8)

   $$RCH=CR'' \xrightarrow{HF} RCH_2-\underset{F}{\underset{|}{\overset{R'}{\overset{|}{C}}}}R''$$

   Introduction of F by addition reactions.

   $$RCH=CHR'' \xrightarrow{Pb(OAc)_4-HF} R\underset{F}{\underset{|}{C}}H-\underset{F}{\underset{|}{C}}HR'$$

REACTIONS OF ORGANIC HALIDES

1. Elimination of hydrogen halide (5.3.2, Pb. 5.10)

$$R-\underset{X}{\underset{|}{C}}H-\overset{H}{\overset{|}{C}}H-R \xrightarrow[C_2H_5OH]{KOH} R-CH=CH-R' + HX$$

$$\xrightarrow{KOH} KX + H_2O$$

$$R-\underset{X}{\underset{|}{\overset{H}{\overset{|}{C}}}}-\underset{X}{\underset{|}{\overset{H}{\overset{|}{C}}}}-R' \xrightarrow{NaNH_2} R\underset{X}{\underset{|}{C}}=\overset{H}{\overset{|}{C}}R' \xrightarrow{NaNH_2} RC{\equiv}CR'$$

11

A valuable method for forming carbon-carbon multiple bonds; the reverse reaction of addition of HX.

2. Substitution of H for X; hydrocarbons form halides (4.3.2, Pb. 9.11)

$$RX \xrightarrow[\text{Zn-HCl, HOAc}]{[H]} RH$$

$Ac = CH_3\underset{\underset{O}{\|}}{C}$, HOAc

Employed in the preparation of saturated hydrocarbons; the reverse of the halogenation of saturated hydrocarbons.

3. Substitution of halogen by a nucleophile (Section 9.2)

$:N^{\ominus}$ = nucleophile

$RX \xrightarrow{:N^{\ominus}} RN + :X^{\ominus}$

$RX + :\ddot{O}H_2 \longrightarrow ROH + HX$

$RX + :NH_3 \longrightarrow RNH_2 \cdot HX$

In these substitution reactions, the nucleophile may be negatively charged or neutral; in either case a nonbonding electron pair is available. The reaction involves an S_N2 or (and) an S_N1 mechanism.

4. Aromatic nucleophilic substitution (9.13, 9.14)

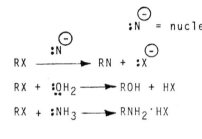

M = <u>meta</u>-directing group

The reaction requires an m-directing group in the p- and (or) o- position to halogen or positions 1 or 4 to a heterocyclic ring N.

5. Organometallic reagents (Section 9.3)

a. Formation (9.15)

$RX + M \longrightarrow RM + MX$
$ArX + M \longrightarrow ArMX$

M = metal

Halides react with metals to give a carbon-metal bond. With Mg metal, the Grignard reagent is obtained.

b. Reaction of Grignard reagents with water (9.16)

$RMgX + HOH \longrightarrow RH + MgX(OH)$

Overall, this gives a saturated hydrocarbon from a halide as the starting point.

PREPARATION OF ALCOHOLS, PHENOLS, AND THIOLS

1. Substitution of a halogen atom by OH (10.2.1)

    a. $RX + :OH^- \longrightarrow ROH + :X^-$

    $X = Cl, Br, I$

    Elimination is a competing reaction particularly with 3° halides but may be suppressed by the use of a weak nucleophile such as water.

    b. $ArX + :OH^- \longrightarrow ArOH + :X^-$

    Because of the lack of reactivity of aryl halides, severe conditions are required.

    c. $RX + SH^- \longrightarrow RSH + X^-$

    A nucleophilic substitution reaction analogous to 1a above.

2. Hydration of alkenes (10.2.2)

    $$RCH=CR'' \xrightarrow[\text{2. HOH}]{\text{1. }HOSO_2OH} RCH_2CR''(R')(OH)$$

    This is the reverse reaction of the formation of C=C by dehydration and results in Markownikoff addition.

3. Hydroboration (10.2.3)

    $$3RCH=CR'' \xrightarrow[B_2H_6]{[BH_3]} \left[B-CH(R)-CHR''(R') \right]_3 \xrightarrow[H_2O_2]{NaOH} RCH(R')-CHR''(OH)$$

    Anti-Markownikoff addition to an alkene by the same mechanism as hydration; attachment of B marks the position of the OH group.

4. Fusion of sodium sulfonates (10.2.4)

    $$ArH \xrightarrow{HOSO_2OH} ArSO_2OH \xrightarrow{NaOH}$$

    $$ArSO_2ONa \xrightarrow[\substack{\text{heat} \\ \text{2. } H^+ - H_2O}]{\text{1. NaOH (solid)}} ArOH$$

REACTIONS OF ALCOHOLS AND PHENOLS

1. Formation of salts (10.3.1)

   $$ROH \xrightarrow{M} ROM + H_2\uparrow$$

   $$ArOH \xrightarrow{NaOH} ArONa + H_2O$$

   Alcohols are less acidic than phenols in that they react only with active metals and not with sodium hydroxide.

2. Formation of esters with inorganic esters (10.3.2)

   $$ROH + HA \longrightarrow RA + H_2O$$

   A = inorganic anion

   Alcohols but not phenols react with inorganic acids to give inorganic esters; phenols require inorganic acid chlorides.

3. Formation of halides (10.3.3, 9.1.2)

   $$ROH + HX \longrightarrow RX + H_2O$$

   X = Cl, Br, I

   This reaction is used as a method of preparation of halides.

4. Electrophilic substitution (10.3.4)

   $$\text{C}_6\text{H}_5\text{OH} + 3X_2 \longrightarrow \text{2,4,6-}X_3\text{C}_6\text{H}_2\text{OH} + 3HX$$

   X = Cl, Br

   Although chlorination can be controlled to give dichloro substitution, the usual reaction is complete substitution in the p- and o- positions with halogen. Multiple substitution is more difficult with $-NO_2$ or $-SO_3H$ groups.

PREPARATION OF ETHERS AND THIOETHERS

1. Symmetrical ethers from alcohols (10.4.1)

   $$ROH + HOR \xrightarrow[\text{heat}]{H_2SO_4} R-O-R + H_2O$$

   This reaction is in competition with elimination to an alkene but can be promoted by keeping the temperature below that optimum for elimination.

2. Unsymmetrical ethers by the Williamson synthesis (10.4.2)

   $RX + NaOR' \longrightarrow R-O-R' + NaX$ This method is unsatisfactory for 3°
   $X = Cl, Br, I$ halides since elimination occurs.

3. Unsymmetrical aryl ethers through a benzyne intermediate (10.4.3)

   $ArX + KOC(CH_3)_3 \longrightarrow Ar-O-C(CH_3)_3 + KX$

   A 3° alkoxide is needed to furnish a sufficiently strong base.

4. Thioethers

   a. Symmetrical (10.23)

   $2RX \xrightarrow{Na_2S} R-S-R + NaX$

   b. Unsymmetrical (10.23)

   $RX + NaSR' \longrightarrow R-S-R' + NaX$ As in other substitution reactions 3° halides must be avoided.

REACTIONS OF ETHERS

1. Cleavage of ethers (10.19, Pb. 10.16)

   $R-O-R' \xrightarrow{HI} RI + R'I + H_2O$ The hydrolytic cleavage of ethers is the main reaction of interest for this series. An aryl group gives a phenol rather than a halide.

   $Ar-O\text{-}R \xrightarrow{HI} ArOH + RI + H_2O$

2. Cleavage of an epoxide ring (10.22)

   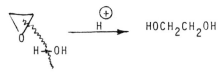 $\xrightarrow{H^+} HOCH_2CH_2OH$ Because of the instability of a 3-membered ring, this type of ether is cleaved far more easily.

PREPARATION OF AMINES

1. Reaction of alkyl halides with ammonia or amines (11.2.1)

$$RX + NH_3 \longrightarrow RNH_2 \cdot HX \xrightarrow{NaOH} RNH_2$$

$$X = Cl, Br, I$$

$$RX + R'NH_2 \longrightarrow RNHR' \cdot HX$$

$$RX + \begin{matrix} R' \\ R'' \end{matrix}\!\!NH \longrightarrow RN\begin{matrix} R' \\ R'' \end{matrix} \cdot HX$$

As has been noted for other nucleophilic substitution reactions, the reaction does not proceed well with 3° halides because of the competing elimination reaction.

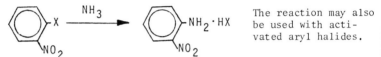

The reaction may also be used with activated aryl halides.

2. Reduction of nitriles (11.2.2)

$$RX \xrightarrow{NaCN} RC\equiv N \xrightarrow{H_2,\ Ni} RCH_2NH_2$$

This reaction may be compared with the hydrogenation of C-C multiple bonds. The overall route is from a halide.

3. Reduction of nitro compounds (11.2.3)

$$ArH \xrightarrow[H_2SO_4]{HONO_2} ArNO_2 \xrightarrow{[H]} ArNH_2$$

This is the common approach to the preparation of aromatic 1° amines.

REACTIONS OF AMINES

1. Salt formation (11.1)

   $RNH_2 + HX \longrightarrow RNH_2 \cdot HX$

   X = anion for any strong acid

   [pyridine] + HX $\longrightarrow$ [pyridinium] · HX

   Salt formation occurs for any 1°, 2°, or 3° amine or when the 2° or 3° amino group is part of a ring.

2. Formation of quaternary ammonium salts (11.3.1)

   $RNH_2 + R'X \longrightarrow RNHR' \cdot HX \xrightarrow{R'X} \underset{R'}{\overset{R'}{RN \cdot HX}} \xrightarrow{R'X} \underset{R'}{\overset{R'}{R-N-R'}}{}^{(+)}X^{(-)}$

   A quaternary ammonium salt can be formed by reaction of an alkyl halide with any type of amine.

3. Formation of amides and sulfonamides (11.3.2)

   $RNH_2 + R'COCl \longrightarrow RNHCOR' + HCl$

   $R_2NH + R'SO_2Cl \longrightarrow R_2NSO_2R' + HCl$

   1° or 2° amines will react with acid chlorides or sulfonyl chlorides to yields amides or sulfonamides, respectively.

4. Electrophilic substitution of aromatic amines (11.3.3)

   PhNH$_2$ + X_2 $\longrightarrow$ 2,4,6-trihalo-aniline   X = Cl, Br

   The -NH$_2$, -NHR, or -NR$_2$ groups activate the benzene ring so strongly that multiple substitution of the para and ortho positions occurs. Other electrophiles can also react with aromatic amines, but in these cases, multiple substitution may not occur for various reasons.

5. Reaction of amines with nitrous acid (11.3.4)

   a. Primary

   $$RNH_2 \xrightarrow{HONO} ROH, RCl, \text{alkenes} + N_2\uparrow$$

   $$ArNH_2 \xrightarrow[NaNO_2-HCl]{HONO} ArN_2^{(+)} Cl^{(-)}$$

   Nitrous acid reacts with aliphatic 1° amines to give nitrogen and a variety of organic products including those resulting from rearrangement (Pb. 11.17). Reaction of nitrous acid with aromatic amines gives relatively stable diazonium salt solutions.

   b. Secondary

   $$\underset{R'}{\overset{R}{>}}NH + HONO \xrightarrow{NaNO_2-HCl} \underset{R'}{\overset{R}{>}}N-NO + HOH$$

   Nitrous acid reacts to give yellow N-nitrosamines. This is a characteristic test for 2° amines but caution must be exercised because aromatic N-nitrosamines are potent carcinogens.

   c. Tertiary

   $$R_3N + HONO \longrightarrow R_3N \cdot HONO$$

   $$R_2N-\langle\bigcirc\rangle-H + HONO \longrightarrow R_2N-\langle\bigcirc\rangle-NO + HOH$$

   Aliphatic 3° amines usually give an aqueous solution of the amine salt, while 3° amines involving an aromatic nucleus undergo substitution of a nitroso group in the para position, the product being green in color.

REACTIONS OF DIAZONIUM SALTS

1. Heating of the diazonium salt solution (11.4.1)

   $ArN_2^{(+)} X^{(-)} \xrightarrow[\text{heat}]{HOH} ArOH + N_2 + HX$

   X = Cl, HSO_4 frequently; also Br

   > This reaction is a synthetic method used to introduce a phenolic hydroxyl group.

2. Introduction of halogen, the Sandmeyer reaction (11.4.2)

   $ArN_2^{(+)} X^{(-)} \xrightarrow{CuX} ArX + N_2$

   X = Cl, Br

   $ArN_2^{(+)} I^{(-)} \xrightarrow{KI} ArI + N_2$

   > The Sandmeyer reaction which employs cuprous halide is a valuable synthetic procedure for introduction of Cl or Br into an aromatic nucleus. For the introduction of I, the cuprous salt is unnecessary. To avoid product mixtures, the acid corresponding to the halogen being introduced often is used to generate the nitrous acid.

3. Introduction of a nitrile group (11.4.3)

   $ArN_2^{(+)} X^{(-)} \xrightarrow{CuCN} ArCN \longrightarrow ArCO_2H$

   > This extension of the Sandmeyer reaction provides an excellent route to the CN group, and ultimately the CO_2H group.

4. Coupling (11.4.4)

   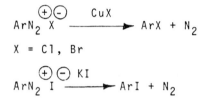

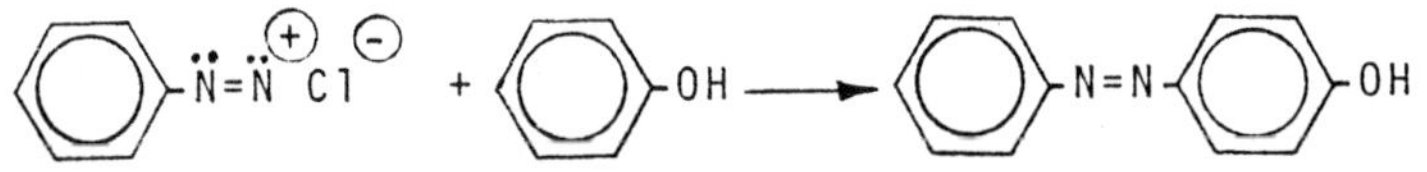

   > A reaction of diazonium salts of great practical value in the manufacture of azo dyes.

PREPARATION OF ALDEHYDES

1. Dehydrogenation of primary alcohols (12.1.1)

$$RCH_2OH \xrightarrow[300°]{Cu} RCHO + H_2$$

This catalytic procedure for the preparation of aldehydes is useful for both laboratory and commercial scale.

2. Friedel-Crafts reaction (12.1.2)

$$ArH + \overset{H}{FC}{=}O \xrightarrow{AlCl_3} ArCHO$$

This is a useful procedure for the preparation of aromatic aldehydes.

3. Reduction of acid chlorides (12.1.3)

$$RCOCl \xrightarrow[LiAlH(OC_4H_9\text{-}\underline{t})_3]{[H]} RCHO$$

Other functional groups, such as the NO_2 group, are not reduced under the same conditions.

4. Oxo process (12.10)

$$RCH{=}CH_2 + CO + H_2 \xrightarrow[150°\text{, pressure}]{[Co(CO)_4]_2} RCH_2CH_2CHO + R\overset{CH_3}{\overset{|}{C}}HCHO$$

The straight chain aldehyde can be made the predominant product.

PREPARATION OF KETONES

1. Oxidation or dehydrogenation of secondary alcohols (12.1.1)

$$\begin{matrix}R\\R'\end{matrix}\!\!>\!CHOH + [O] \xrightarrow[Na_2Cr_2O_7\text{-}H_2SO_4]{} \begin{matrix}R\\R'\end{matrix}\!\!>\!C{=}O + H_2O$$

Alicyclic 2° alcohols can also be converted to alicyclic ketones. Dehydration with copper catalyst is used as well as chemical oxidizing agents.

2. Pyrolysis of carboxylic acid salts (12.1.4)

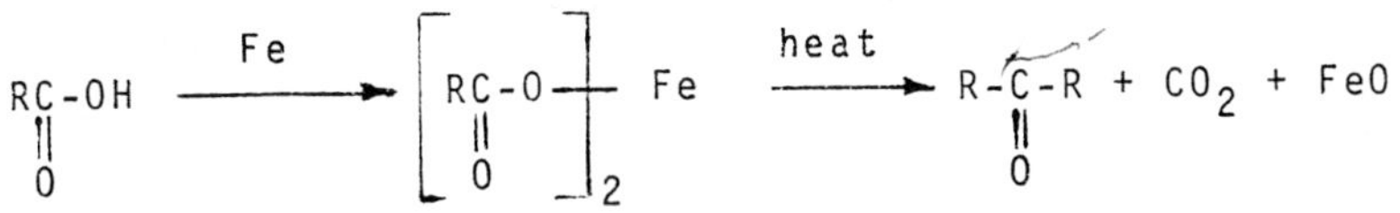

$$RC(=O)OH \xrightarrow{Fe} [RC(=O)O]_2 Fe \xrightarrow{heat} R-\overset{O}{\underset{\|}{C}}-R + CO_2 + FeO$$

This is an excellent, general laboratory procedure for the synthesis of ketones where the organic groups are identical.

A UNIQUE REACTION OF ALDEHYDES

1. Oxidation (12.3.1)

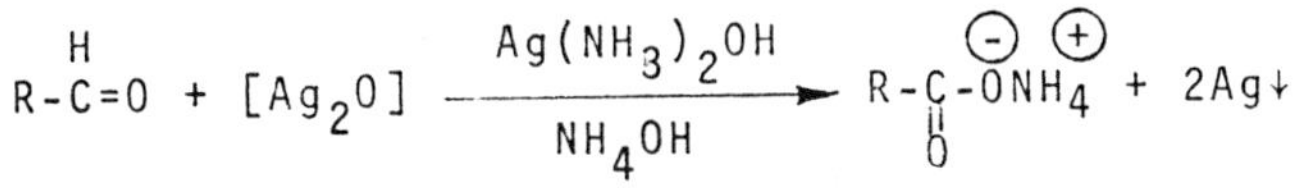

$$R-\overset{H}{\underset{}{C}}=O + [Ag_2O] \xrightarrow[NH_4OH]{Ag(NH_3)_2OH} R-\overset{O}{\underset{\|}{C}}-\overset{\ominus}{O}\overset{\oplus}{NH_4} + 2Ag\downarrow$$

Oxidation, employing Tollens' reagent, is a unique property of aldehydes, and is used as a test. Similarly, Benedict's or Fehling's tests are also used (Pb. 12.4).

ADDITION REACTIONS OF ALDEHYDES AND KETONES

1. Reduction of the carbonyl group (12.3.2)

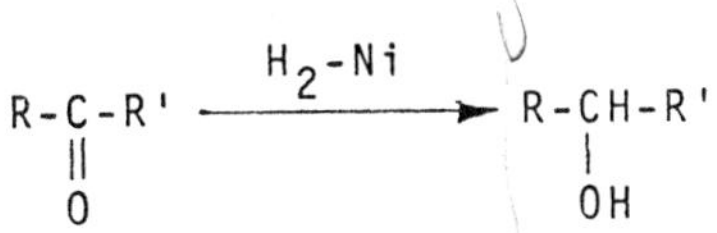

$$R-\overset{O}{\underset{\|}{C}}-R' \xrightarrow{H_2-Ni} R-\overset{}{\underset{OH}{C}}H-R'$$

Reduction of a carbonyl is the reverse reaction to dehydrogenation used as a method of preparation. The general formula represents an aldehyde where R' = H. Reactions throughout are common to both aldehydes and ketones unless otherwise stated.

2. Addition of hydrogen cyanide (12.3.3)

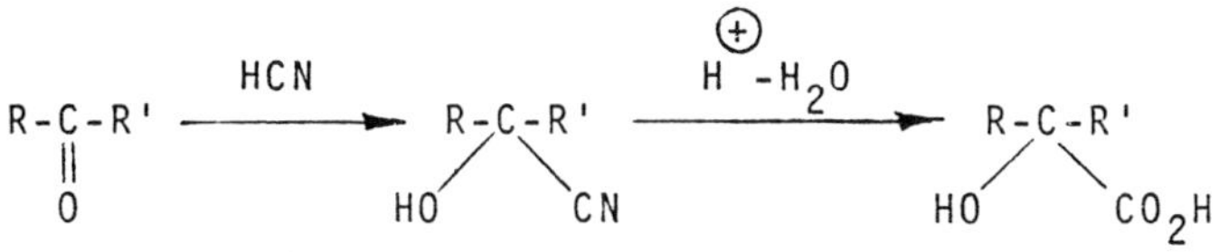

$$R-\overset{O}{\underset{\|}{C}}-R' \xrightarrow{HCN} R-\underset{HO\quad CN}{C}-R' \xrightarrow{H^{\oplus} \; -H_2O} R-\underset{HO\quad CO_2H}{C}-R'$$

The addition of hydrogen cyanide to form a cyanohydrin, which is a useful intermediate in the synthesis of hydroxy acids.

3. Reaction with Grignard reagents (12.3.4)

$$R\text{-}\underset{\underset{O}{\|}}{C}\text{-}R' \xrightarrow[(2)\ H^{\oplus}\text{-}H_2O]{(1)\ R''MgX} R\text{-}\underset{\underset{OH}{|}}{\overset{\overset{R''}{|}}{C}}\text{-}R' + MgX(OH)$$

Where $R = R' = H$ (formaldehyde), 1° alcohols are obtained; where R = organic group, $R' = H$ (aldehydes in general), 2° alcohols; and where $R = R'$ = organic groups (ketones), 3° alcohols.

4. Addition of alcohols (12.3.6)

$$R\text{-}\underset{\underset{O}{\|}}{C}\text{-}R' + HOR'' \xrightarrow{H^{\oplus}} \left[\underset{HO\ \ \ \ OR''}{\overset{R\text{-}C\text{-}R'}{\diagup\diagdown}}\right] \xrightarrow[H^{\oplus}]{HOR''} \underset{R''O\ \ \ \ OR''}{\overset{R\text{-}C\text{-}R'}{\diagup\diagdown}}$$

a hemiacetal   an acetal

This reaction only occurs readily for $R' = H$ (aldehydes) because in the case of most ketones the equilibrium is unfavorable. The ketone counterpart (ketals) can be made by another method, however.

5. Addition of nitrogen-containing groups (12.3.6)

$$R\text{-}\underset{\underset{O}{\|}}{C}\text{-}R' + H_2NG \longrightarrow \left[\underset{HO\ \ \ \ NHG}{\overset{R\text{-}C\text{-}R'}{\diagup\diagdown}}\right] \longrightarrow R\text{-}\underset{\underset{NG}{\|}}{C}\text{-}R'$$

G = -NH⟨phenyl⟩, -NH⟨phenyl⟩-NO$_2$, -NHCONH$_2$, or -OH

The initial addition is followed by an elimination reaction to give the product which is often a useful derivative in characterizing aldehydes and ketones.

6. Addition of reactive methyl and methylene groups (12.3.7)

$$RCH_2CR' + RCH_2CR' \xrightarrow{\overset{\ominus}{OH}} \underset{R}{\underset{|}{HO}}\overset{RCH_2CR'}{\underset{CHCOR'}{\diagdown}} \xrightarrow[\text{heat}]{\overset{\oplus}{H}\ -H_2O} \underset{\underset{RCCOR'}{\parallel}}{RCH_2CR'}$$
$$\underset{O}{\parallel} \quad \underset{O}{\parallel}$$

                              Unlike the analogous N situation above, it is possible often to isolate the addition product. Heating with acid, however, gives the elimination product.

7. The haloform reaction (12.3.8)

$$R-\underset{OH}{\underset{|}{CH}}-CH_3 \xrightarrow{NaOX} R-\underset{\underset{O}{\parallel}}{C}-CH_3 \xrightarrow{NaOX} R-\underset{\underset{O}{\parallel}}{C}-CX_3 \longrightarrow R-\underset{\underset{O}{\parallel}}{C}-ONa + HCX_3$$

                       $+ H_2O$ $+ NaOH$

X = Cl, Br, I

                              Where X = I, this reaction is used as a useful test to identify substances, principally alcohols and methyl ketones, which conform to the general formula given.

8. Cannizzaro reaction (12.3.11)

    R-CHO + R-CHO + NaOH $\longrightarrow$ R-CH$_2$OH + R-COONa

                              Where R = an organic group with no α-hydrogen (e.g., C_6H_5); certain 1° alcohols and acids are prepared by this oxidation-reduction reaction.

PREPARATION OF CARBOXYLIC ACIDS

1. Oxidation of alcohols or aldehydes (15.3.1)

$$RCH_2OH \xrightarrow{[O]} RCO_2H \xleftarrow{[O]} RCHO$$

                              This is a useful method of generating the -CO$_2$H group providing there are no other functional groups in the molecule which are sensitive to oxidation.

2. Oxidation of side chains (8.12, 15.3.2)

$$Ar-R \xrightarrow{[O]} Ar-CO_2H$$

The side chain R may be of any length and also contain an easily oxidized functional group. The aromatic ring Ar may be benzene or a heterocyclic ring such as pyridine (Reaction 15.8). The Ar nucleus may have up to six R side chains attached where Ar is benzene, all of which are oxidatively degraded in the reaction to $-CO_2H$ groups.

3. Hydrolysis of nitrile groups (15.3.3)

$$RX \xrightarrow{NaCN} RCN \xrightarrow{H^{+} -H_2O} RCO_2H$$

Alkyl halides, RX where R is 1° or 2°, can be converted to nitriles, and the nitrile group then can be hydrolyzed to a $-CO_2H$ group.

4. Reaction of a Grignard reagent with carbon dioxide (15.3.4)

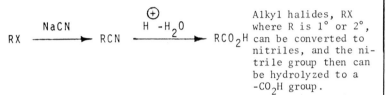

This synthetic scheme is particularly useful where R is 3° and, hence, a -CN group cannot be introduced because the elimination reaction predominates. Or, if R is an aromatic nucleus, this route through a Grignard intermediate is available.

REACTIONS OF CARBOXYLIC ACIDS

1. Salt formation (15.4.1)

$$RCO_2H \underset{HCl}{\overset{NaOH}{\rightleftarrows}} RCO_2^{-} \; Na^{+}$$

Salt formation is often a useful reaction in the preparation and separation of acids. Free acids are obtained from their salts by the reverse reaction, acidification.

2. Acid chloride formation (15.4.3)

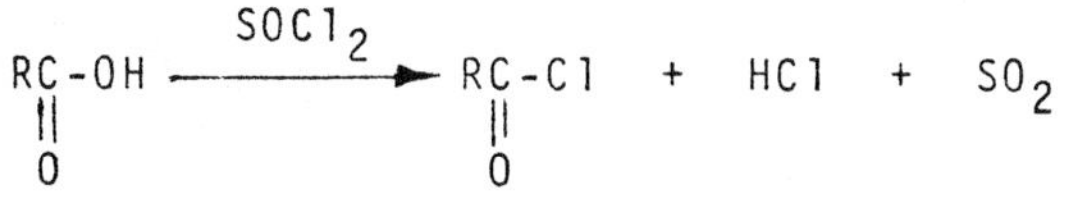

$$RC(=O)\text{-OH} \xrightarrow{SOCl_2} RC(=O)\text{-Cl} + HCl + SO_2$$

This reaction is important in preparing acid chlorides, reactive acid intermediates employed in making other acid derivatives, particularly amides and esters.

3. Acid anhydride formation (15.4.2)

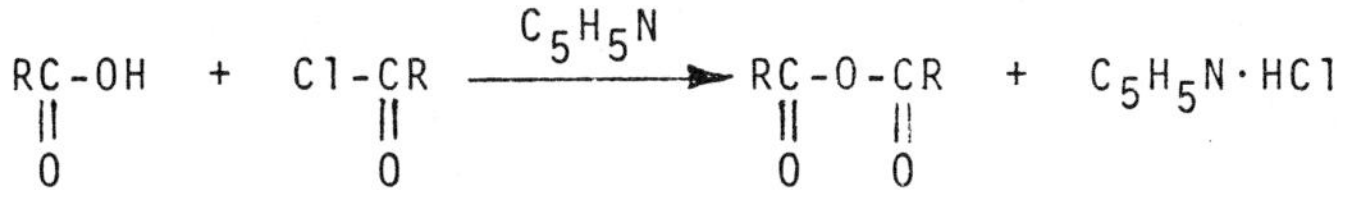

$$RC(=O)\text{-OH} + Cl\text{-}CR(=O) \xrightarrow{C_5H_5N} RC(=O)\text{-O-}CR(=O) + C_5H_5N\cdot HCl$$

The R groups need not necessarily be equivalent but usually are because these are the useful anhydrides in preparing acid derivatives. A mixed anhydride structure would lead to mixtures of products.

4. Ester formation (15.4.4)

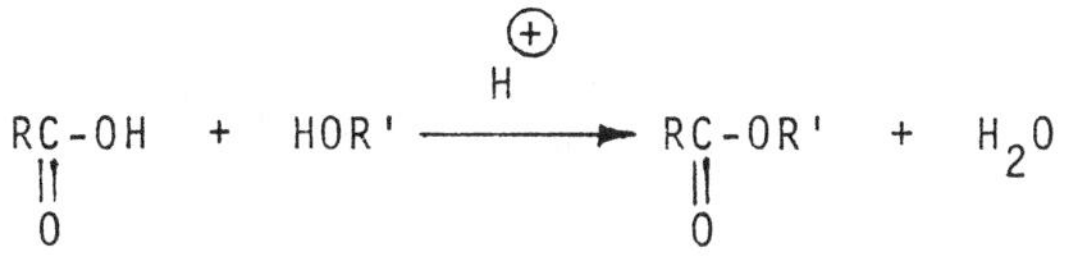

$$RC(=O)\text{-OH} + HOR' \xrightarrow{H^{\oplus}} RC(=O)\text{-OR'} + H_2O$$

Some esters are made industrially by this direct method involving the acid itself, others indirectly through the acid anhydride. Esters are very important as consumer products ranging from a textile fiber, polyester, to a medicinal agent, aspirin.

5. Lithium aluminum hydride reduction (15.4.5)

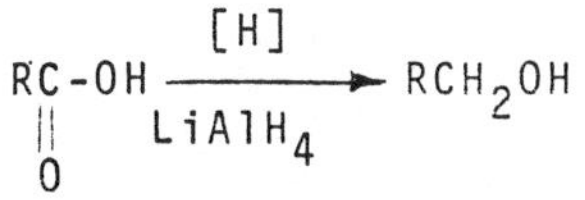

$$RC(=O)\text{-OH} \xrightarrow[LiAlH_4]{[H]} RCH_2OH$$

$LiAlH_4$ is one of the few reagents capable of reducing a $-CO_2H$; however, because of its relatively high cost, its use on an industrial scale is limited to the production of expensive chemicals such as medicinal agents.

6. α-Hydrogen substitution (15.4.6)

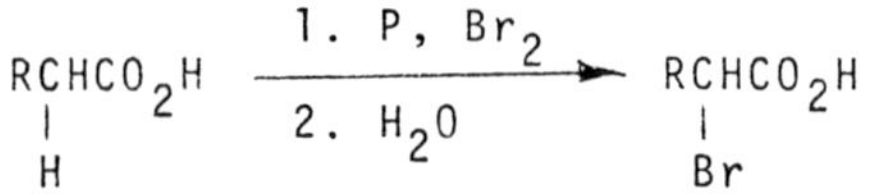

$$\underset{H}{\overset{|}{RCHCO_2H}} \xrightarrow[2.\ H_2O]{1.\ P,\ Br_2} \underset{Br}{\overset{|}{RCHCO_2H}}$$

H atoms on a C atom attached to a $\underset{O}{\overset{\|}{C}}$ of the -CO_2H group are more subject to substitution than unactivated C-H bonds, particularly through the intermediacy of an acid halide.

7. Aromatic substitution (15.4.7)

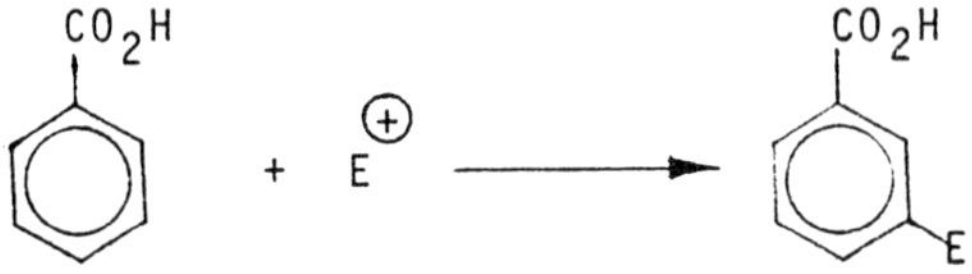

Ph-CO_2H + E⁺ ⟶ m-E-C₆H₄-CO_2H

The $\underset{O}{\overset{\|}{C}}$ of the -CO_2H group also has the effect of being meta-directing in aromatic electrophilic substitution.

PREPARATION OF ESTERS

1. Direct from the carboxylic acid and alcohol (15.4.4, 15.5.1). See item 4 in Review Box on reactions of carboxylic acids.

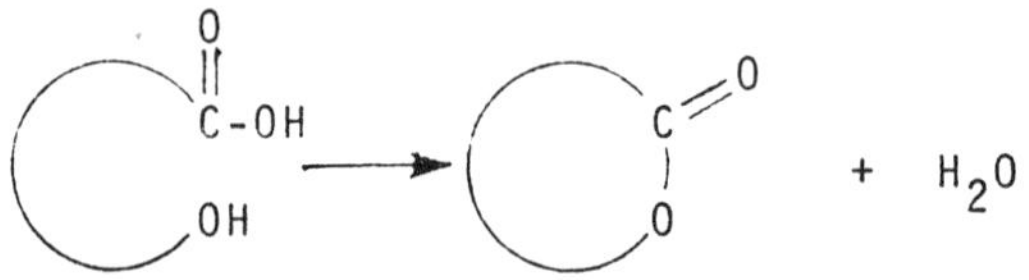

[Ring]-C(=O)-OH + [Ring]-OH ⟶ lactone (C=O, O in ring) + H_2O

Intramolecular esterification occurs readily to form 5- and 6-membered rings known as lactones.

2. From acid anhydrides (15.5.2)

$$\underset{O}{\overset{\|}{RC}}-O-\underset{O}{\overset{\|}{CR}} + HOR' \longrightarrow \underset{O}{\overset{\|}{RC}}-O-R' + HO-\underset{O}{\overset{\|}{CR}}$$

Esters are frequently made through the intermediacy of the more reactive acid anhydride derivative.

3. From acid chlorides (15.5.3)

$$RC\text{-}Cl + HOAr \longrightarrow RC\text{-}OAr + HCl$$
$$\|\|$$
$$O\phantom{\text{-}Cl + HOAr \longrightarrow RC}O$$

Acid chlorides are particularly useful in preparing esters of phenols and <u>tert</u>-alcohols, as derivatives for identification.

REACTIONS OF ESTERS

1. With water (15.6.1)

$$R\text{-}C\text{-}OR' + HOH \xrightarrow{NaOH} R\text{-}C\text{-}ONa + HOR'$$
$$\phantom{R\text{-}}\|\phantom{\text{-}OR' + HOH \xrightarrow{NaOH} R\text{-}}\|$$
$$\phantom{R\text{-}}O\phantom{\text{-}OR' + HOH \xrightarrow{NaOH} R\text{-}}O$$

The most useful hydrolysis reaction of esters is under alkaline conditions and is known as saponification. Hydrolysis under physiological conditions occurs in the process of digestion of fats.

2. With alcohols (15.6.2)

$$R\text{-}C\text{-}OR' + HOR'' \xrightarrow{NaOR''} R\text{-}C\text{-}OR'' + HOR'$$
$$\phantom{R\text{-}}\|\phantom{\text{-}OR' + HOR'' \xrightarrow{NaOR''} R\text{-}}\|$$
$$\phantom{R\text{-}}O\phantom{\text{-}OR' + HOR'' \xrightarrow{NaOR''} R\text{-}}O$$

This reaction is known as transesterification and has a practical application in the production of polyester synthetic textile fibers.

3. With ammonia or primary or secondary amines (15.6.3)

$$R\text{-}C\text{-}OR' + H_2NR'' \longrightarrow R\text{-}C\text{-}NHR'' + HOR'$$
$$\phantom{R\text{-}}\|\phantom{\text{-}OR' + H_2NR'' \longrightarrow R\text{-}}\|$$
$$\phantom{R\text{-}}O\phantom{\text{-}OR' + H_2NR'' \longrightarrow R\text{-}}O$$

This is a reaction useful in preparing unsubstituted or substituted amides, sometimes on a commercial scale.

4. Reduction of esters (15.6.4)

   a.
   $$RC(=O)-OR' \xrightarrow[\text{LiAlH}_4]{[H]} RCH_2OH + HOR'$$

   > This is a useful reaction in synthesis. A C=C in the R group will not be reduced.

5. With Grignard reagents (15.6.5)

   $$RC(=O)-OR' + R''MgX \xrightarrow[\text{H}^+ \; -H_2O]{\text{Followed by}} R-\underset{OH}{\overset{R''}{C}}-R''$$

   > This reaction constitutes a synthetic method for the preparation of 3° alcohols, except that in the case of formate esters, 2° alcohols with <u>two identical</u> R groups are obtained. Carbonate esters give 3° alcohols <u>with three identical</u> R groups.

PREPARATION OF ACID CHLORIDES

1. Reaction of acids with thionyl chloride (15.4.3)

   $$RC(=O)-OH \xrightarrow{SOCl_2} RC(=O)-Cl + HCl + SO_2$$

   > This is a method of choice for preparing acid chlorides. See item 2 in Review Box on reactions of carboxylic acids.

REACTIONS OF ACID CHLORIDES

1. With water (15.7.1)

   $$RC(=O)-Cl + HOH \longrightarrow RC(=O)-OH + HCl$$

   Acid chlorides react vigorously with water, which is often a detrimental side reaction in the use of acid chlorides in synthesis.

2. With alcohols or phenols (15.26)

$$RC\text{-}Cl + HOR' \longrightarrow RC\text{-}OR' + HCl$$
$$\|\|$$
$$OO$$

> This reaction is often used as a method of making solid esters as derivatives of alcohols and phenols. See item 3 in Review Box on preparation of esters.

3. With carboxylic acids (15.16)

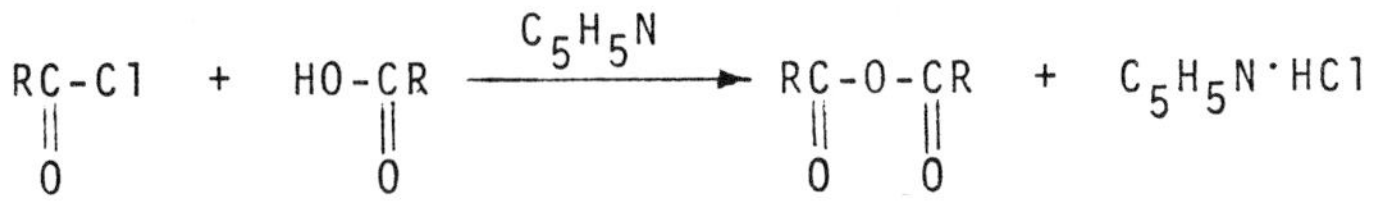

$$RC\text{-}Cl + HO\text{-}CR \xrightarrow{C_5H_5N} RC\text{-}O\text{-}CR + C_5H_5N\cdot HCl$$
$$\|\|\phantom{\xrightarrow{C_5H_5N} RC}\|\phantom{\text{-}O\text{-}}\|$$
$$OO\phantom{\xrightarrow{C_5H_5N} RC}O\phantom{\text{-}O\text{-}}O$$

> Note the similarity of this reaction with that in item 2 involving alcohols or phenols.

4. With ammonia or primary or secondary amines (15.7.2)

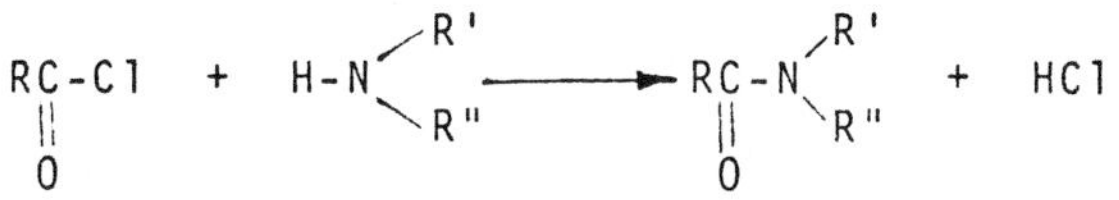

$$RC\text{-}Cl + H\text{-}N\begin{matrix}R'\\R''\end{matrix} \longrightarrow RC\text{-}N\begin{matrix}R'\\R''\end{matrix} + HCl$$
$$\|\phantom{+ H\text{-}N\begin{matrix}R'\\R''\end{matrix} \longrightarrow RC}\|$$
$$O\phantom{+ H\text{-}N\begin{matrix}R'\\R''\end{matrix} \longrightarrow RC}O$$

> The reaction is illustrated with a 2° amine which forms an N, N-disubstituted amide. In like manner, a 1° amine gives an N-monosubstituted amide and ammonia an unsubstituted amide. Compare item 3 in Review Box on reactions of esters.

5. With arenes to form aldehydes or ketones (12.1.2, Scheme 8.10)

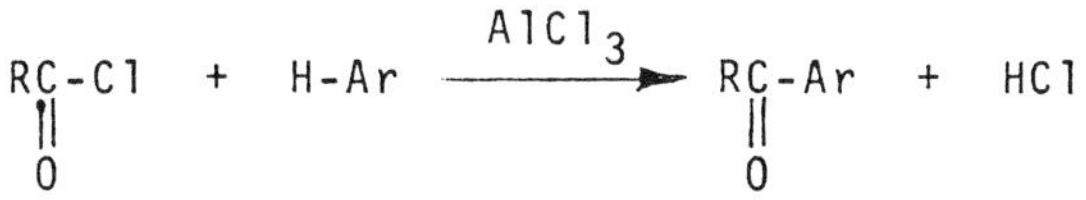

$$RC\text{-}Cl + H\text{-}Ar \xrightarrow{AlCl_3} RC\text{-}Ar + HCl$$
$$\|\phantom{+ H\text{-}Ar \xrightarrow{AlCl_3} RC}\|$$
$$O\phantom{+ H\text{-}Ar \xrightarrow{AlCl_3} RC}O$$

> This type of reaction is widely used for the preparation of aldehydes and ketones.

PREPARATION OF ACID ANHYDRIDES

1. By the intramolecular interaction of carboxyl groups to form 5- and 6-membered ring anhydrides (15.39)

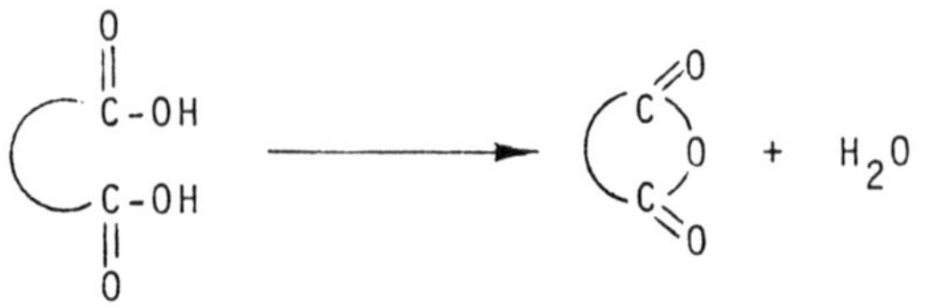

$$\begin{array}{c} \text{C-OH} \\ \text{C-OH} \end{array} \quad \longrightarrow \quad \begin{array}{c} \text{C} \\ \text{O} \\ \text{C} \end{array} \quad + \quad H_2O$$

Carboxyl groups do not ordinarily interact to form an anhydride, but they do so easily if the reaction is intramolecular to form a 5- or 6-membered ring. For the general method for the preparation of anhydrides, see item 3 in the Review Box on reaction of acid chlorides.

REACTIONS OF ACID ANHYDRIDES

1. With water, alcohols and phenols, and ammonia and amines, and arenes, respectively (Section 15.9)

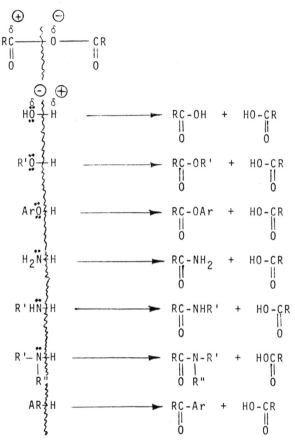

This comparative scheme shows the similarity of all of these nucleophilic reactions. The desired change is substitution of an acyl group (R—C—) for H in each in-
$\|$
O

stance. Note also the formation of an equivalent of carboxylic acid. Similar schemes should be written out for acid chlorides and esters (except for the Friedel-Crafts reaction in the latter case) and the three schemes compared among themselves.

PREPARATION OF AMIDES

1. From esters (15.30), acid chlorides (15.35), and acid anhydrides (15.42)

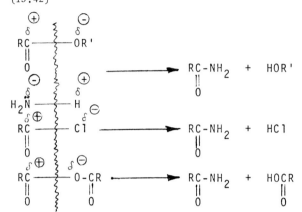

    This comparative scheme again provides a useful summary. A summary of the formation of substituted amides from 1° and 2° amines, respectively, can be written out in the same way.

2. From ammonium salts of carboxylic acids (Section 15.10)

$$RC(=O)-OH + NH_3 \longrightarrow RC(=O)-O^- \; NH_4^+ \xrightarrow{heat} RC(=O)-NH_2 + H_2O$$

    This reaction is particularly useful in the preparation of amides on an industrial scale.

AN IMPORTANT REACTION OF AMIDES

1. With water (15.11.1)

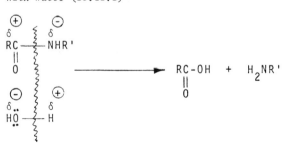

Although amides are less reactive than other acid derivatives, hydrolysis is a very important reaction. It is the reaction which involves the digestion of proteins to amino acids so that they can be utilized in building body tissue. The above reaction is a model example of this biochemical change, and note that it is the reverse of amide formation. Unsubstituted and $\underline{N}$, $\underline{N}$-disubstituted amides undergo hydrolysis as well.

PREPARATION OF NITRILES

1. Dehydration of amides (15.11.2)

$$RC(=O)-NH_2 \xrightarrow[\text{heat}]{P_2O_5} \left[\begin{array}{c} RC=NH \\ | \\ OH \end{array} \right] \longrightarrow RC\equiv N + H_2O$$

This reaction may be viewed as the elimination of water from the enol tautomer of the amide. The reaction is particularly useful where R is 3°, as the substitution reaction listed next fails in this case.

2. By nucleophilic substitution (9.8)

$$RX + :\overset{\ominus}{C}N \longrightarrow RCN + :\overset{\ominus}{X}$$

This reaction is effective where R is 1° but fails where R is 3° because of the competing elimination reaction.

REACTIONS OF NITRILES

1. Hydrolysis (15.13.1)

$$RC\equiv N + HOH \xrightarrow{\overset{\oplus}{H}} \left[\begin{array}{c} RC=NH \\ | \\ OH \end{array} \right] \longrightarrow$$

$$\text{RC-NH}_2 + \text{HOH} \xrightarrow{\overset{\oplus}{H}} \text{RC-OH} + \text{NH}_3$$
$$\underset{\text{O}}{\|} \qquad\qquad\qquad \underset{\text{O}}{\|}$$

Note that this reaction is the reverse of dehydration of an amide, item 1 above in the Review Box on preparation of nitriles.

2. Reduction (15.13.2)

$$\text{RC} \equiv \text{N} \xrightarrow{\text{H}_2-\text{Ni}} \text{RCH}_2\text{NH}_2$$

This reaction has commercial significance in that it is used for the production of 1,6-hexanediamine which, in turn, is used in the production of nylon-66.

ANSWERS TO PROBLEMS

In this section the answers to all the problems in the text are presented. It is intended that these will provide further instruction in the subject and will expand the coverage of the text as some solutions furnish supplementary information. Comments on how the problems are approached are frequently provided. It is recommended that a solution in this section should not be looked at until after some thought has been given to solving the problem.

chapter 1

BACKGROUND AND INTRODUCTORY CONCEPTS

1. (1) $H_3C\text{-}\boxed{C\text{-}}CH_3$ (with $\|$ O below, circled) (2) $H_2\boxed{C=CH}\boxed{CH_2}\text{-}Cl$ (3) $\boxed{HO\text{-}CH_2\text{-}CHCH_2\text{-}OH}$ with $\overset{|}{OH}$

   (4) $\boxed{HS}CH_2CH_2\boxed{NH_2}$

2. If the total percentage composition given does not total 100%, the difference is assumed to be %O; 39.8 + 6.68 = 46.48, 100 - 46.48 = 53.52%. The % of each element is then divided by the atomic weight of that element; 39.8 ÷ 12.01 = 3.32; 6.68 ÷ 1.01 = 6.68; 53.5 ÷ 16 = 3.34. Each of these ratios is then divided by the smallest ratio: C, 3.32 ÷ 3.32 = 1, H, 6.68 ÷ 3.32 = 2.02; O, 3.34 ÷ 3.32 = 1. The empirical formula, therefore, is CH_2O. Since the empirical formula gives a formula weight of 30, dividing the molecular weight by 30 (90.1 ÷ 30) gives a multiple of 3. Thus, $3(CH_2O)$ gives a molecular formula of $C_3H_6O_3$. The substance is (+)-lactic acid.

3. 35.6 + 2.39 + 33.3 = 71.29   100 - 71.29 = 28.71 %O
   C, 35.6/12 = 2.97, 2.97/1.79 = 1.66, 1.66 x 3 ∿ 5
   H, 2.39/1 = 2.39, 2.39/1.79 = 1.33, 1.33 x 3 ∿ 4
   N, 33.3/14 = 2.39, 2.39/1.79 = 1.33, 1.33 x 3 ∿ 4
   O, 28.7/16 = 1.79, 1.79/1.79 = 1, 1 x 3 ∿ 3

   Since atoms are present in molecules in a ratio of whole numbers, it is necessary to multiply fractional ratios by a common factor which gives a ratio of whole numbers. In this case, the factor is 3. Thus, the empirical formula is $C_5H_4N_4O_3$, and since this equals MW 168, it is also the molecular formula. The substance is uric acid.

4. 53.9 + 4.77 + 7.89 + 9.00 + 6.45 + 81.98 = 82.21
   100.00 - 82.21 = 17.79 %O

   C, 53.9/12 = 4.50, 4.50/0.281 = 16 $C_{16}H_{17}N_2O_4SNa$
   H, 4.77/1 = 4.77, 4.77/0.281 = 17
   N, 7.89/14 = 0.561, 0.561/0.281 = 2   MW, 356
   O, 18.0/16 = 1.12, 1.12/0.281 = 4
   S, 8.99/32 = 0.289, 0.289/0.281 = 1   sodium salt of
   Na, 6.46/23 = 0.281, 0.281/0.281 = 1   penicillin G

5. (1) C_3H_6O (2) C_3H_5Cl (3) $C_3H_8O_3$ (4) C_2H_7NS

6. A reactive or functional group either involves a double bond (two valence bonds joining the atoms) and/or an element other than carbon. Another multiple bond also possible in functional groups is a triple bond.

(1) Nylon 6-6

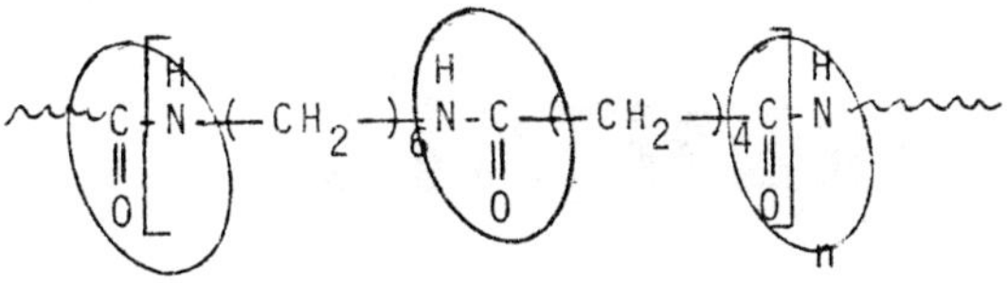

(2) $C_6H_5-CH=CH-C=O$ with H — Cinnamaldehyde, a constituent of oil of cinnamon.

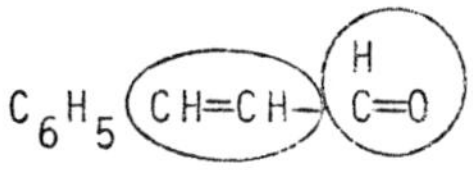

(3) Active component in the oral contraceptive, Provest.

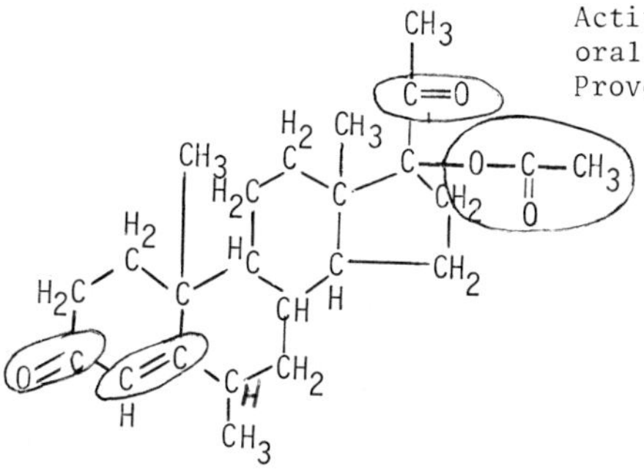

(4) Cocaine, an alkaloid from the leaves of the coca plant, is a dangerous drug which causes addiction. It induces euphoria and alleviates fatigue. It also has anesthetic properties, and some valuable synthetic anesthetics, used in medical practice, have been based on its structure.

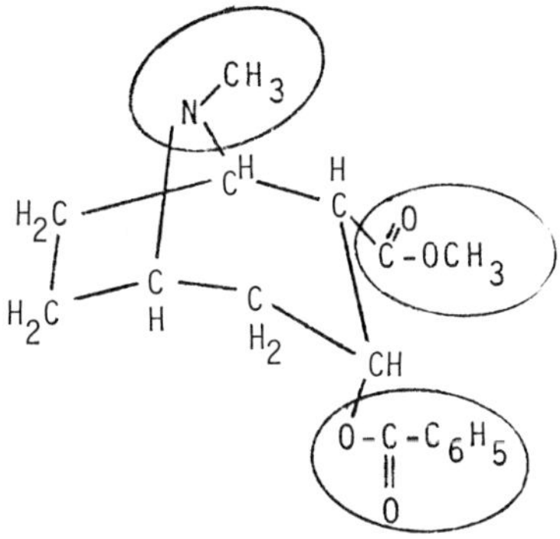

(5) Chloramphenicol, an antibiotic useful in the treatment of typhoid fever. *The double bonds in a benzene ring do not behave as an isolated $C=C$ which is explained in the text book (pp. 38, 149, 154, 155). Therefore, in this instance, they are not considered a functional group.*

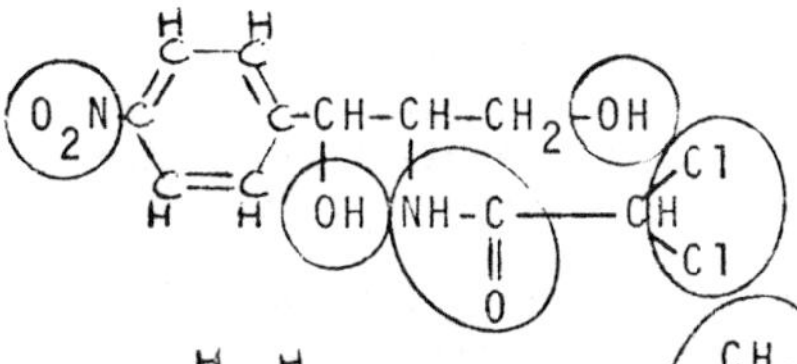

(6) Anahist, an antihistaminic agent.

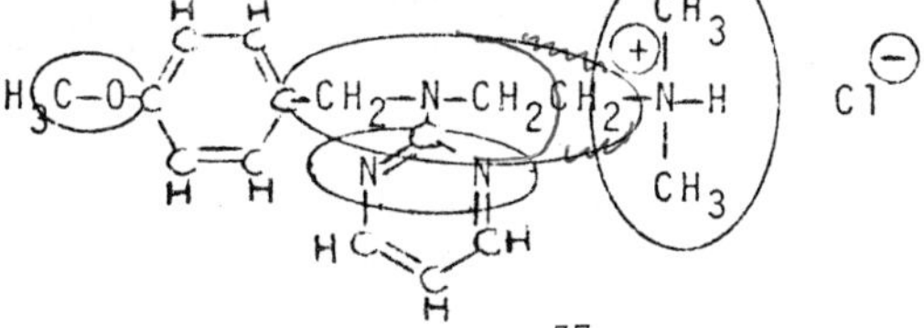

37

chapter 2

VALENCE AND FUNDAMENTALS OF STRUCTURE

1. 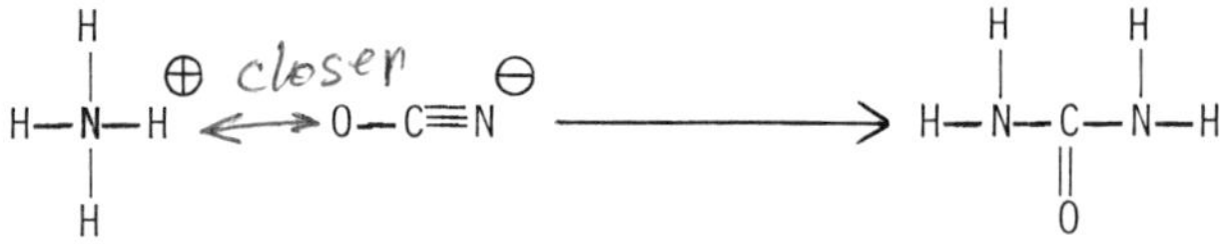 H—N(H)(H)—H $\oplus$ *closer* $\ominus$ O—C≡N ⟶ H—N(H)—C(=O)—N(H)—H

2. (1) $CH_3CH_2CHClCH_2CH(OH)CH_2Cl$ (2) $CH_3(CH_2)_3CH_3$
   (3) $(CH_3)_3CI$ (4) $(NC)_2C=C(CN)_2$

3. (1) 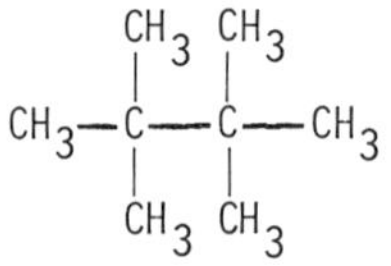
$$CH_3-\underset{\underset{CH_3}{|}}{\overset{\overset{CH_3}{|}}{C}}-\underset{\underset{CH_3}{|}}{\overset{\overset{CH_3}{|}}{C}}-CH_3$$

   The center bond is elongated to prevent the methyl substituents from giving a crowded appearance.

   (2) 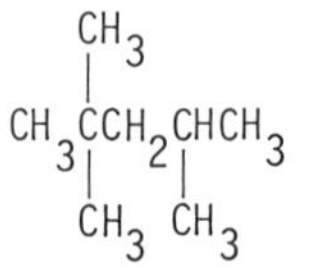
$$CH_3\underset{\underset{CH_3}{|}}{\overset{\overset{CH_3}{|}}{C}}CH_2\underset{\underset{CH_3}{|}}{\overset{\overset{}{|}}{C}}HCH_3$$

   (3)
$$CH_3CH=CHC(=O)CH(CH_3)CH_3$$

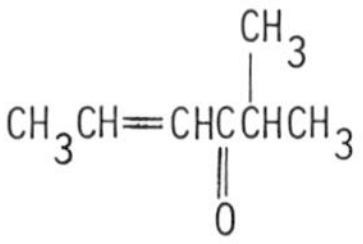

   Multiple bonds involving carbon and other atoms are often extended, C=O. Substituents are positioned above the chain as well as below to prevent crowding.

   (4) 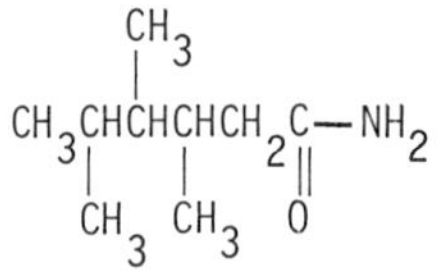
$$CH_3CH(CH_3)CHCH(CH_3)CH_2C(=O)NH_2$$

4. (1) I-Cl: from Table 2.1, 3.0 - 2.5 = 0.5; from Table 2.2, % ionic character, 6.5%.   (2) C-Mg, Mg-I: The difference in the electronegativities for either of these bonds is the same. 2.5 - 1.2 = 1.3; from Fig. 2.1, 1.3 corresponds to about 34%.   (3) P-Br: 2.8 - 2.1 = 0.7; from Fig. 2.1 about 12.5%.   (4) H-F: 4.0 - 2.1 = 1.9; from Fig. 2.1 about 58%.   (5) Al-Cl: 3.0 - 1.5 = 1.5; about 42%.   (6) N-H: 3.0 - 2.1 = 0.9; about 18.5%.

5. The more stable arrangement, actually, places one of the tetrahedra in an inverted position in respect to the other. This gives what is known as a staggered arrangement (or conformation) of the hydrogen atoms (6.13) (Fig. A-2.5).

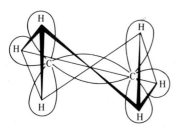

Figure A-2.5

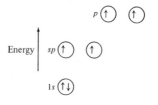

Figure A-2.6

6. Another possibility follows logically from sp^3 and sp^2 hybridization; it is $sp^{(1)}$ (or just always noted as sp) hybridization (Fig. A-2.6).

problem 7.

7. The hybridization scheme described in Pb. 6 requires that sp orbitals have an angle of 180° for maximum separation of orbitals. The sp orbital overlapping with another sp orbital or an orbital of another atom give molecular orbitals corresponding to σ bonds. Also for maximum separation, the p orbitals must be perpendicular to themselves and to the sp orbitals. When the sp orbitals of two C atoms are so joined, the two adjacent sets of p orbitals coalesce to give two π orbitals (Fig. A-2.7).

sp-sp not sp

It is indicated by some authors that the p orbitals coalesce to give a cylindrical electron cloud (Fig. A-2.7). This hydrocarbon is acetylene (ethyne) (2.3).

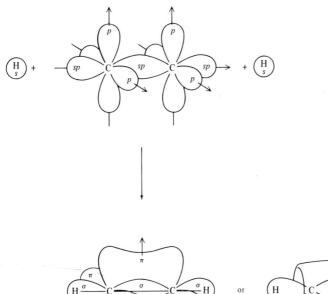

Figure A-2.7

The two π orbitals in the left hand fig. can coalesce to give a cylindrical orbital about the sigma orbital as an axis.

8.

9. Join two N atoms together with oxygen using the number of valence electrons available, five for N and six for O. Then consider other bonding alternatives and decide which structures would be most likely on chemical grounds.

   5 5 6

   (1)   (2)   (3)

   Note that the terminal N in structure (1) lacks an octet of electrons. All atoms have completed octets, however, in both structures (2) and (3). Thus, the structure of nitrous oxide is a resonance hybrid to which (2) and (3) contribute. Aside form the octet requirement, structures with like charges on adjacent atoms would be unstable. Thus, structure (4) is unimportant for this reason:

   $^{2\ominus}$ ⊕ ⊕
   :N̈—N≡O:

   (4)

10. ⊕   ⊖   ⊖   ⊕
    H—F ↔ H   :F ↔ H: F

    (1)   (2)   (3)

   Structures (1) and (2) are important, whereas (3) is of no importance. Fluorine is a highly electronegative element and thus can readily take on an extra election as in (2), but not give one up as in (3).

   ⊕   ⊖   ⊖   ⊕
   H_3C :H ↔ H_3C—H   ↔ H_3C: H

   (1)   (2)   (3)

   Unlike HF, where the differing electronegativities of the elements results in considerable ionic character as represented by (2), the similar electronegativities of C and H lead to a highly covalent character, and thus the ionic species (1) and (3) are unimportant resonance structures. This type of structure, however, is of importance as a highly reactive and transient species in organic reactions (4.12).

11. (1) H—Ö—H + H$^{⊕}$ → H—Ö—H
                                d |
                                  H

   Both the electrons of this nonbonding pair used to form the bond a with H$^{⊕}$. However, all three of the bonds are equivalent and partially coordinate covalent.

(2)

$$\text{H}-\text{O}-\overset{\overset{:\overset{\ominus}{\text{O}}:}{\underset{\underset{:\overset{..}{\text{O}}:}{b}}{a|2\oplus}}}{\text{S}}-\text{O}-\text{H}$$

The S—O bonds marked a and b are equivalent and each is formed by S contributing both the electrons of the pair to form the bond. However, S can expand its valence shell to 10 or 12, and thus S—O is often written as S=O.

(3)

$$\text{H}-\underset{\underset{..}{|}}{\text{N}}-\text{H} + \text{H}^{\oplus} \rightarrow \text{H}-\overset{\overset{\text{H}}{|}}{\underset{\underset{\text{H}}{|}}{\text{N}^{\oplus}}}-\text{H}$$
$$ a$$

In parallel to (2), N contributes both the electrons of the pair to form N—H bond a.

(4)

$$\text{H}-\text{O}-\underset{a}{\overset{\oplus}{\text{N}}}\underset{:\overset{..}{\text{O}}:^{\ominus}}{\overset{\overset{..}{\text{O}}.}{\diagup\diagdown}}$$

In the N—O bond a, both of the electrons are contributed by N to form the bond.

12. (1) $\text{Cl}-\underset{\underset{\text{Cl}}{|}}{\text{Al}}-\text{Cl}$

The Al is two electrons short of an octet and is therefore electron-attracting and a Lewis acid.

(2) $\text{H}-\overset{..}{\underset{..}{\text{O}}}-\text{H}$

Two nonbonding electron pairs; thus, it is electron-donating to form a bond [see Pb. 2.11(1) above], Lewis base.

(3) $\text{CH}_3^{\oplus}$

An electron pair short of an octet, a Lewis acid (4.12).

(4) $:\overset{\ominus}{\underset{..}{\text{N}}}\text{H}_2$

Electron pair of this anion readily available for bonding, a strong Lewis base.

(5) $(\text{C}_6\text{H}_5)_3\overset{\ominus}{\text{C}}:$

Electron pair of a carbanion, a strong Lewis base (4.12).

(6) BCl_3

B, like Al in (1), is short an electron pair, a Lewis acid. Remember that B has only three valence electrons and with one from each of three Cl's has a total of six, still two electrons short of an octet.

13. The following schemes show how two empty 3d orbitals are employed to give six atomic orbitals of S for overlap with six orbitals of a strongly electron-attracting element such as F (Fig. A-2.13).

Maximum separation of six sp^3d^2 atomic orbitals gives an arrangement wherein the orbitals are directed toward the corners of a regular octahedron (Fig. A-2.13). Overlap with fluorine gives molecular orbitals with the same geometric shape.

3p (↑↓)(↑)(↑) sp^3d^2 (↑)(↑)(↑)(↑)(↑)(↑)
3s (↑↓)
2p (↑↓)(↑↓)(↑↓)   2p (↑↓)(↑↓)(↑↓)
2s (↑↓)   2s (↑↓)

1s (↑↓)   1s (↑↓)

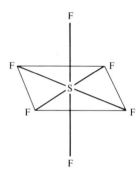

Figure A-2.13

14. (1) Covalent bond: A bond formed by approximately equal sharing of an electron pair by like or different elements of closely similar electronegativities. In ethane,

```
    H H
    | |
H—C—C—H
    | |
    H H
```

either the C—C or C—H bonds are covalent.

(2) sp^2 Hybridization: Orbitals for carbon, which are one part s and two parts p, with one p orbital remaining. For maximum separation, an angle of 120° is required for the three sp^2 orbitals; the p orbital is perpendicular to this plane (Fig. 2.12).

(3) Polarized bond: A covalent bond involving an unequal sharing of a pair of electrons:

$\overset{\delta\oplus\delta\ominus}{H-Cl}$

(4) Condensed structural formula: A formula where all hydrogen atoms and groups are accumulated with the total number of each indicated by the proper subscript. It can be written on a single line:

$(CH_3)_3CCH_2CH(C_2H_5)CH_2CH_3$

(5) Line formula for a cyclic compound: A line drawing of the appropriate geometric shape, wherein each corner represents a C and two H atoms (unless substituted or involved in a double bond):

(6) Functional group: Functional groups are those that readily undergo a reaction in an organic molecule (functional groups are circled in the example):

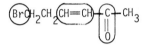

(7) π-Orbital: A π-orbital is a molecular orbital formed by the overlap of p orbitals on adjacent atoms in conjunction with sp^2 or sp hybridization (Fig. 2.12).

(8) Sigma bond: A sigma bond results for carbon by overlap with sp^3, sp^2, or sp between C atoms. Overlap may occur with H and other elements as well (see appropriate molecular orbital drawings in this chapter).

(9) Molecular orbital: A molecular orbital results from the overlap of atomic orbitals.

(10) Extended structural formula: An extended structural formula requires each individual atom in the molecule be written out:

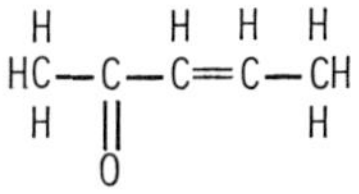

```
    H     H H H
    HC—C—C=C—CH
    H   ||    H
        O
```

(11) Lewis acid: A Lewis acid is any charged or neutral species which is electron-attracting:

$AlCl_3$, $CH_3^{\oplus}$

(12) sp^3 Atomic orbital: A hybridized atomic orbital containing one part s character and three parts p. In combination with orbitals of other atoms, it forms molecular orbitals.

Maximum separation requires that these orbitals be directed toward the corners of a regular tegrahedron (Fig. 2.8).

(13) Line formulas for straight-chain compounds are drawn in a zig-zag fashion with the angle approximating the normal C—C bond angle of 109°28'. The end of each line and each corner represents a C atom with the appropriate number of H atoms attached:

$CH_3(CH_2)_7CH=CH(CH_2)_7CO_2H$

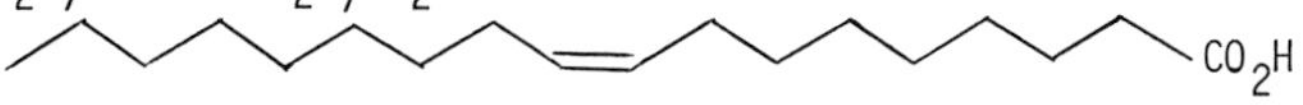

(14) Coordinate covalent bond: A bond in which both electrons of the pair are furnished by one atom, rather than one electron by each atom:

$$H-O-\overset{\oplus}{\underset{\underset{O_{\ominus}}{|a}}{S}}-O-H$$

The S—O bond at a is coordinate covalent in sulfurous acid.

15. [Note in parts (3) and (6) that x's and o's are helpful in keeping account of the valence electrons from each atom. Nonbonding electron pairs are shown in formulas only when instructional clarity is desired. Electron bookkeeping with x's and o's can be used with any example.]

(1) $H_2\overset{\oplus}{C}=\overset{\ominus}{N}=N:$ ↔ $H_2\overset{\ominus}{C}-\overset{\oplus}{N}\equiv N:$

(2) $O=\overset{\oplus}{\underset{\underset{O_{\ominus}}{|}}{P}}-O-\overset{\oplus}{\underset{\underset{O_{\ominus}}{|}}{P}}=O$

(3) $:\overset{\ominus}{C}\equiv\overset{\oplus}{O}:$, $\overset{x}{_x}C\overset{x}{_x}\overset{o}{_o}O\overset{o}{_o}$

(4) $H_3C-N\overset{\diagup O}{\underset{\overset{\oplus}{O}_{\ominus}}{\diagdown}}$

(5) $\overset{\ominus}{\underset{\underset{O_{\ominus}}{|}}{\overset{O}{\underset{|}{\ominus O-\overset{3\oplus}{Cl}-O-H}}}}$

(6) $H-O-N=O$, $H\overset{x}{_o}O\overset{o}{_x}\overset{+}{N}\overset{o o}{\underset{o o}{\diagup O\overset{o}{_o}}}\overset{o o}{\diagdown O\overset{o}{_o}}$

(7) $H_3C-\overset{\overset{O_{\ominus}}{|}}{\underset{\underset{O_{\ominus}}{|}}{\overset{|2\oplus}{S}}}-CH_3$

(8) $KO-\overset{\overset{OK}{|}}{\underset{\underset{O}{||}}{P}}-OK$

(9) $F-\overset{..}{\underset{\diagup\diagdown}{S}}-F$ $F\quad F$

(10) $O=C=O$

16. $NH_4^{\oplus}$ shows sp^3 hybridization of N and because there are two electrons in each orbital totaling an octet, the species is stable. For $CH_3^{\oplus}$, there is also sp^3 hybridization but one of the orbitals is empty. Since there are only six electrons, or a deficit of two, the species is highly unstable.

17. In BF_3, sp^2 hybridization is likely with the three molecular orbitals in a plain 120° apart. In acquiring a share in the pair from nitrogen, boron would become sp^3 hybridized.

18. $O=C=C=C=O$

chapter 3
CLASSES OF ORGANIC COMPOUNDS AND NOMENCLATURE

1. (1) C_8H_{18} (2) $C_{11}H_{24}$ (3) $C_{17}H_{36}$ (4) $C_{20}H_{42}$
   (5) $C_{35}H_{72}$

2. (1) $CH_3(CH_2)_6-$, heptyl (2) $CH_3(CH_2)_2-$, propyl
   (3) $CH_3(CH_2)_{10}CH_2-$, dodecyl (4) $CH_3(CH_2)_7-$, octyl
   (5) $CH_3(CH_2)_{18}-$, nonadecyl

3. (1) $(CH_3)_2CHCH_2CH_3$, isopentane, $(CH_3)_2CH(CH_2)_2-$, isopentyl
   (2) $(CH_3)_2CH(CH_2)_2CH_3$, isohexane, $(CH_3)_2CH(CH_2)_3-$, isohexyl
   (3) $(CH_3)_2CH(CH_2)_3CH_3$, 2-methylhexane, $(CH_3)_2CH(CH_2)_4-$, 5-methylhexyl

4. A systematic method of determining all isomeric possibilities for lower members of a series is as follows: (1) Select a carbon atom bearing the functional group. (2) Attach to this carbon atom every possible combination of simple groups:

   C_5 alkanes, $CH_3CH_2CH_2CH_2CH_3$, $(CH_3)_2CHCH_2CH_3$, $(CH_3)_4C$;

   C_5 alkyl chlorides: carbon with the functional group, C-Cl;
   then attach all the possible C_4 groups, $CH_3CH_2CH_2CH_2CH_2Cl$,
   $(CH_3)_2CHCH_2CH_2Cl$, $CH_3CH_2CHCH_2Cl$, $(CH_3)_3CCH_2Cl$;
                                         | |
                                         CH_3 CH_3

   now C_1 and C_3 groups, $CH_3CHCH_2CH_2CH_3$, $CH_3CHCHCH_3$;
                                             | |
                                             Cl   Cl

   now C_2, $CH_3CH_2CHCH_2CH_3$;
                                  |
                                  Cl

                                CH_3

   now C_1, C_1, C_2, $CH_3CCH_2CH_3$
                                 |
                                 Cl

5. $CH_3-\underset{\underset{Br}{|}}{\overset{\overset{CH_3}{|}}{C}}-CH_2CH_2CH_3$, $CH_3\underset{\underset{Br}{|}}{\overset{\overset{CH_3}{|}}{C}}CH(CH_3)_2$, $CH_3CH_2\underset{\underset{Br}{|}}{\overset{\overset{CH_3}{|}}{C}}CH_2CH_3$

6. (1) $C_6H_5CH_2NH_2$ (2) $(CH_3)_2CHOH$ (3) cyclopentyl–C(=O)–CH$_3$

    (4) $H_2C=CHCH_2-O-CH_2CH=CH_2$ (5) $(CH_3)_2CHCH_2NHCH_3$

7. (1) <u>sec</u>-butyl alcohol, (2) allyl phenyl ketone, (3) N-ethyl-N-methylcyclohexylamine, (4) benzyl isopentyl sulfide, (5) <u>tert</u>-pentyl bromide

8. (1) 2-methylpentane, (2) 2,3-dimethylbutane. If it is difficult to work with condensed structural formulas, extend them:

    $CH_3CH-CHCH_3$
       $\underset{CH_3}{|}$ $\underset{CH_3}{|}$

    (3) 4-propylheptane, (4) 3,3,6,6-tetraethyloctane, (5) 2,5-dimethyl-3-phenylheptane, (6) 2,7,8-trimethyldecane. Note that it is not 3,4,9-trimethyldecane even though, in this case, the total of the numbers is lower than in the first instance. It is the point of first difference, 2 versus 3, that is the deciding factor.

9. Lowest numbers are correlated with alphabetical ordering where possible.

    $Cl-\underset{F}{\overset{F}{C}}-\underset{\underset{Cl}{|}}{\overset{\overset{F}{|}}{C}}-(CF_2)_4CH_2CH_2I$

10. (1) 5-butyl-4-methylundecane, 5-(1-methylbutyl)undecane is incorrect as it does not give a maximum number of substituents, 1 + 4 + 11 = 16; (2) 4-<u>tert</u>-butylheptane, 4 + 7 = 11; (3) 4-ethyl-7-(1,3,3-trimethylbutyl)dodecane, 2 + 7 + 12 = 21.

11. (1) 3,4-dimethyl-2-pentene, (2) 6-methyl-1-heptyne, (3) 3,4-diethyl-3-hexene, (4) 4,7-dipropyl-5-decyne, (5) 2-methyl-1-hepten-6-yne, (6) 2,4-dimethyl-1,3,5-nonatriene

12. (1) 3-methyl-4-nitro-1-butyne, (2) 4,4,4-trifluoro-1-butene, (3) 5-iodo-1-nitro-5-phenyl-2-pentene, (4) 1-bromo-7-chloro-1-heptyne, (5) 2-bromo-5-chlorohexane

13. (1) 1-<u>tert</u>-butyl-4-isopropylcyclooctane, (2) 6-bromo-1,3-dimethylcyclodecane

14. ▷–, cyclopropyl and similarly cyclobutyl, cyclopentyl, cyclohexyl, cycloheptyl, and cyclooctyl

15.
(1) 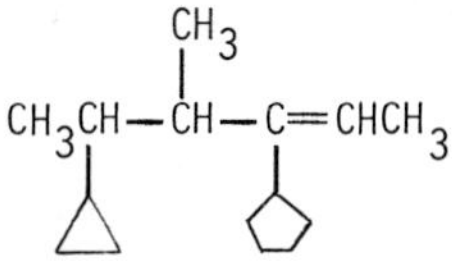 CH₃CH—CH—C=CHCH₃ with CH₃ on second C, cyclopropyl on first CH, cyclopentyl on third C

(2) 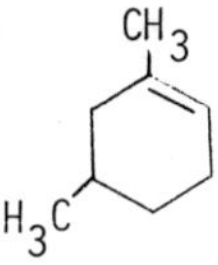 1-methyl-5-methylcyclohexene (structure shown)

(3) 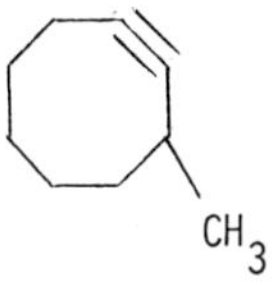 cyclooctyne with CH₃ substituent

(4) 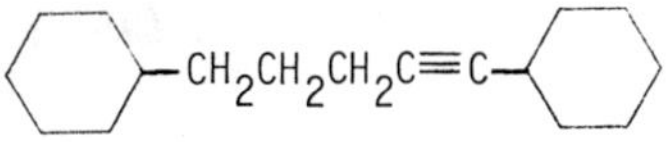 cyclohexyl-CH₂CH₂CH₂C≡C-cyclohexyl

16.
(1) 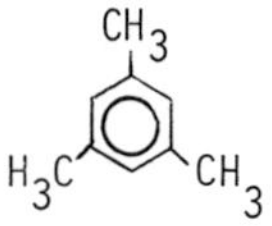 1,3,5-trimethylbenzene (structure with CH₃ groups)
(2) 1,2,4,5-tetramethylbenzene
(3) 1,3-dichloro-5-ethylbenzene
(4) 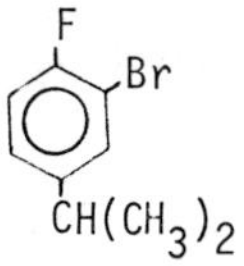 benzene with F, Br, CH(CH₃)₂ substituents

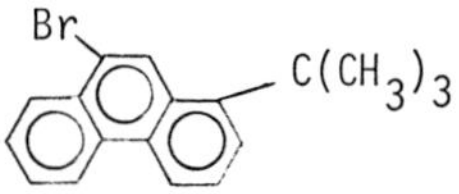

17.
(1) 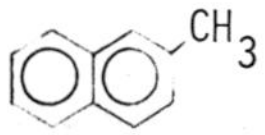 methylnaphthalene (CH₃ substituent)
(2) 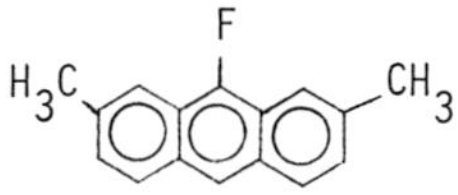 anthracene with F, H₃C, CH₃ substituents
(3) 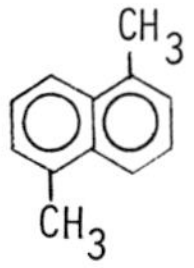 1,5-dimethylnaphthalene
(4) phenanthrene with Br and C(CH₃)₃
(5) anthracene with four C₂H₅ groups

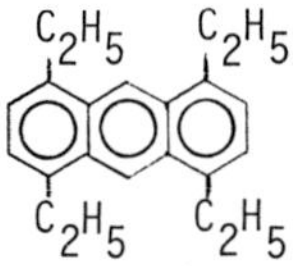

18. (1) 2-butanol, sec-butyl alcohol, 2°; (2) 2-propen-1-ol, allyl alcohol, 1°; (3) 4-methyl-1-pentanol, isohexyl alcohol, 1°; (4) 2-methyl-2-propanol, tert-butyl alcohol, 3°; (5) cyclobutanol, 2°; (6) benzyl alcohol, 1°

19.
(1) 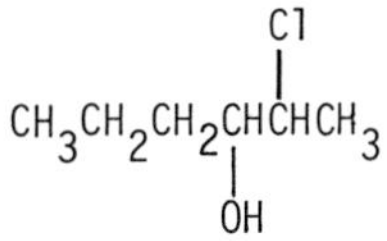 CH₃CH₂CH₂CHCHCH₃ with Cl and OH

(2) 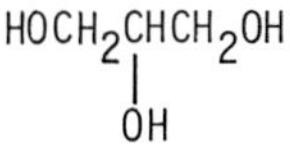 HOCH₂CHCH₂OH with OH

(3) 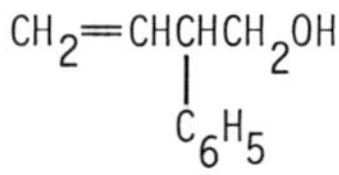 CH₂=CHCHCH₂OH with C₆H₅

(4) 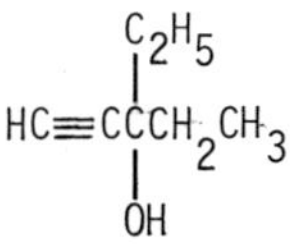 HC≡CCCH₂CH₃ with C₂H₅ and OH

(5) HO(CH₂)₁₀OH

(6) 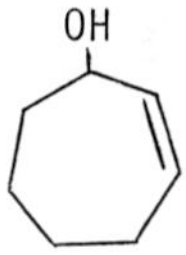 cyclohept-2-en-1-ol

(7) 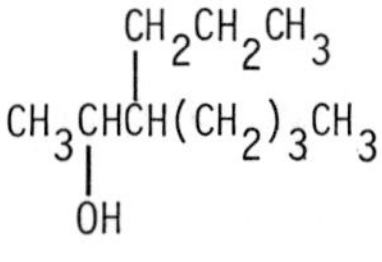 CH₃CHCH(CH₂)₃CH₃ with CH₂CH₂CH₃ and OH

(8) 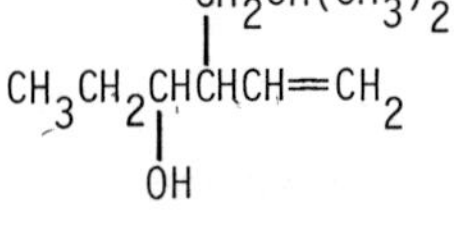 CH₃CH₂CHCHCH=CH₂ with CH₂CH(CH₃)₂ and OH

20. (1) 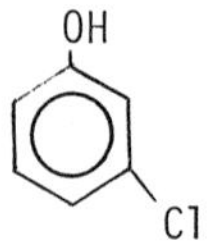 [m-chlorophenol structure: benzene ring with OH and Cl]   (2) p-cresol, p-methylphenol   (3) 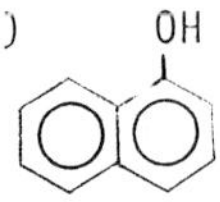 [1-naphthol structure: naphthalene with OH]

(4) 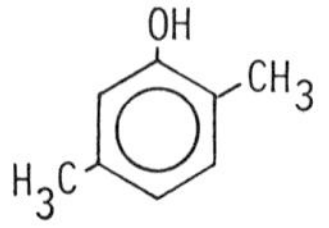 [2,6-dimethylphenol structure: benzene with OH, CH$_3$, and H$_3$C]   (5) 2,6-dibromophenol   (6) p-ethylphenol

21. (1) [(CH$_3$)$_2$CHCH$_2$]$_2$O   (2) diphenyl ether

(3) C$_2$H$_5$OCH(CH$_3$)$_2$ (4) sec-butyl cyclopropyl ether

(5) 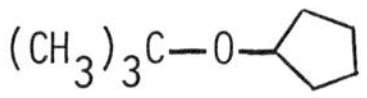 (CH$_3$)$_3$C—O—[cyclopentyl]   (6) cycloheptyl phenyl ether

22. (1) CH$_3$O(CH$_2$)$_4$OH

(2) 1-ethoxy-3-methyl-1-pentyne or ethyl 3-methyl-1-pentynyl ether

(3) 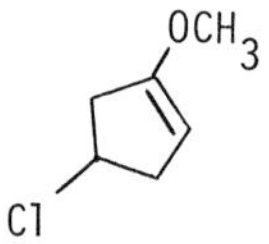 [cyclopentene with OCH$_3$ and Cl substituents]

(4) 1-isopropoxy-3-methoxy-2-phenylpropane

23. (1) CH$_3$CH$_2$CH$_2$SH   (2) CH$_3$CH(CH$_2$)$_3$CH$_3$ with SH

(3) 2-propene-1-thiol   (4) 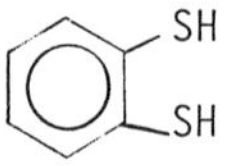 [benzene with two SH groups ortho]

(5) HSCH$_2$CH$_2$CH$_2$CHCH$_3$ with OH

24. (1) C$_6$H$_5$—S—C$_6$H$_5$ (2) isopropyl 4-nitrophenyl sulfide

(3) C$_2$H$_5$—S—CH$_2$CHCH$_2$CH$_3$ with C$_6$H$_5$ (4) 5-(sec-butylthio)-1-pentene

25. (1) C$_2$H$_5$NH$_2$, 1°   (2) (CH$_3$CH$_2$CH$_2$CH$_2$)$_2$NH, 2°

(3) triethylamine, 3°

(4) 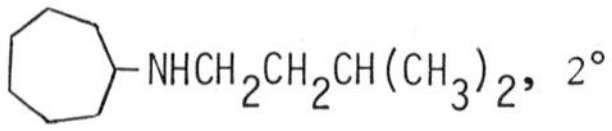 [cycloheptyl]—NHCH$_2$CH$_2$CH(CH$_3$)$_2$, 2°

(5) N-cyclopropyl-N-methylcyclopentylamine, 3°

(6) $CH_3CH_2CH(CH_2)_3OH$, 1° (7) 1-(methylamino)-7-octen-2-ol, 2°
          |
         NH_2

(8) N,N-dimethyl-3-cyclopenten-1-ylamine, 3°

26.
(1) 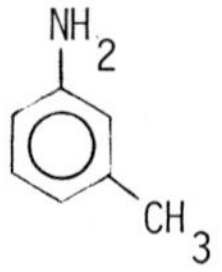 3-methylaniline structure with NH_2 and CH_3 groups (2) 3-chloropyridine structure (3) o-anisidine

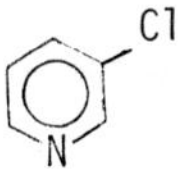

(4) 5-chloro-2-nitropyridine (5) 4,7-dichloroquinoline

(6) 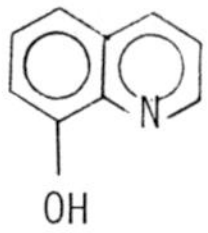 8-hydroxyquinoline structure with OH

27.
(1) $(CH_3)_2CH-C-CCH_2CH_2CH_3$ with =O and I substituents (2) 1-bromo-3-phenyl-2-butanone

(3) $CH_3(CH_2)_4-C-C-CH_2OH$ with $(CH_2)_4CH_3$ branch, OH and =O substituents

(4) 3-oxo-13-phenyltridecanoic acid (5) 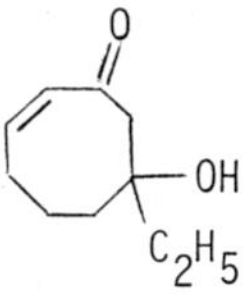 cycloheptanone structure with OH and C_2H_5 substituents

Note that in a ring, the low numbering designates a double bond after the main functional group. Other groups are then assigned the lowest numbers possible and named in alphabetical order.

(6) $CH_3CCH_2CH_2CCH_3$ with two =O groups (7) 2-methyl-2-nonen-4-one

(8) 2-methyl-2-cyclohepten-1-one

28. (1) $(CH_3)_2CHCH_2CH_2CHO$, 4-methylvaleraldehyde
(2) $CH_3(CH_2)_{10}CHO$, dodecanal
(3) lauric acid, dodecanoic acid
(4) $CH_3(CH_2)_6CO_2H$, octanoic acid
(5) caprylaldehyde, octanal
(6) $CH_3(CH_2)_{14}CO_2H$, hexadecanoic acid

(7) palmitaldehyde, hexadecanal   (8) $CH_3(CH_2)_8CHO$, decanal
(9) $CH_3(CH_2)_{16}CO_2H$, octadecanoic acid
(10) stearaldehyde, octadecanal

29. (1) 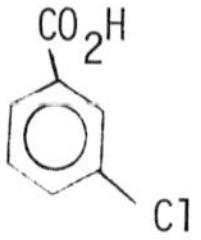 3-chlorobenzoic acid structure with CO_2H and Cl

(2) $CH_3CH_2C{\equiv}CCH_2CH_2CHO$

(3) 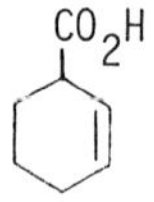 cyclohexenecarboxylic acid structure with CO_2H

(4) 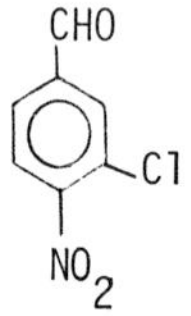 benzaldehyde with Cl and NO_2 substituents

(5) 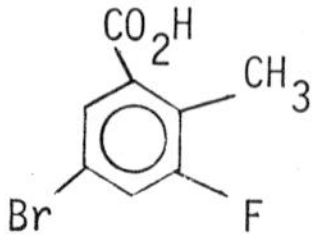 benzoic acid with CO_2H, CH_3, Br, F substituents — also named 5-bromo-3-fluoro-o-toluic acid (CA)

(6) $HO(CH_2)_{15}CHO$

(7) 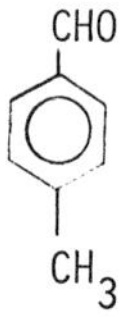 p-methylbenzaldehyde (CHO and CH_3)

30. (1) $CH_3(CH_2)_9\underset{Cl}{CH}CO_2Na$

(2) $C_6H_5CO_2CH_3$

(3) phenyl acetate

(4) magnesium octadecanoate or magnesium stearate (CA)

(5) $(CH_3)_2CHCH_2\underset{O}{\overset{\parallel}{C}}{-}O\underset{CH_3}{CH}CH_2CH_3$, also named sec-butyl isovalerate (CA)

(6) isopropyl diphenylacetate

(7) $CH_3\underset{O}{\overset{\parallel}{C}}OCH_2CH_2O\underset{O}{\overset{\parallel}{C}}CH_3$

also 1,2-ethanediol diacetate or ethylene diacetate

31. (1) 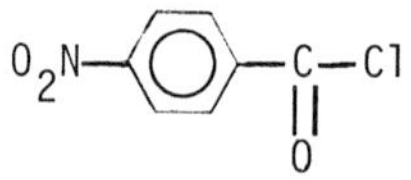 $O_2N{-}C_6H_4{-}\underset{O}{\overset{\parallel}{C}}{-}Cl$

(2) 3,5-dichlorobenzoyl chloride

(3) 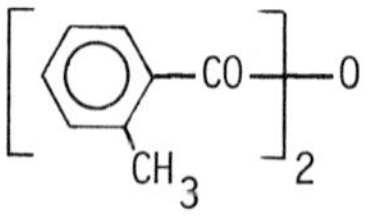 $[C_6H_4(CH_3){-}CO{-}O]_2$

(4) $CH_3(CH_2)_6\underset{CH_3}{CH}CH_2COCl$

(5) $[CH_3(CH_2)_4CO{-}O]_2$, also named hexanoic anhydride (CA)

(6) bis(2-methoxy-3-methylbutyric) anhydride

(7) cyclobutanecarboxylic anhydride

32. (1) $C_6H_5CONH_2$ (2) m-chlorobenzamide

(3) $CH_3CH{=}C(C_2H_5){-}CONH_2$ or 2-ethylcrotonamide (CA);

crotonic acid is $CH_3CH{=}CHCO_2H$

(4) 2-bromo-2-propenamide or 2-bromoacrylamide (CA);

acrylic acid is $CH_2{=}CHCO_2H$

(5) $CH_3CH_2CH(C_2H_5){-}CH(CH_3)CONH_2$, also named 3-ethyl-2-methylvaleramide (CA)

(6) 6,7-dichloro-2-naphthamide (7) $CH_3CH_2CONHC(CH_3)_3$

(8) N-isobutyl-N-methylhexanamide

33. (1) $ClCH_2CH_2CH_2CH_2CN$, also named 5-chlorovaleronitrile (CA)

(2) 2,3-dimethyloctanenitrile

(3) $(CH_3)_3CCN$ (4) 2,2-dimethylpropionitrile

(5) $NC(CH_2)_5CO_2H$ (6) adiponitrile or 1,6-hexanedinitrile

(7) 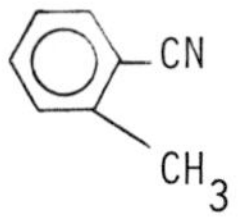

(8) 2,6-dimethylbenzonitrile

34. (1) $CH_3CH(CH_3)CH_2CH(CH_3)CH_2CH(CH_3)CH_3$ (2) 2,2,7,7-tetramethyloctane

(3) $C_6H_5CH_2CH_2CH_2C(CH_2CH_2CH_3){=}CHCH_3$ (4) 3-ethyl-2-heptene

cis or trans

(5) $(CH_3)_2CHCH_2C{\equiv}CC_6H_5$, 4-methyl-1-phenyl-1-pentyne

(6) 1,8-nonadiyne

(7) 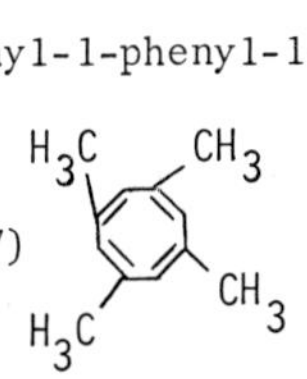

(8) perchlorocyclopentadiene; per indicates all H's substituted by Cl, or 1,2,3,4,5,5-hexachloro-1,3-cyclopentadiene

(9) 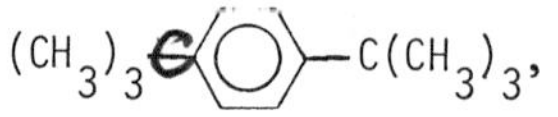 $(CH_3)_3C$-⟨O⟩-$C(CH_3)_3$, also p-di-tert-butylbenzene

(10) 1-sec-butyl-3-cyclobutylbenzene

(11) 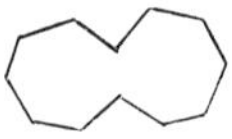 [bicyclic structure]   (12) cyclohexadecyne

35.
(1) $CH_3CH_2\underset{\underset{CH_3}{|}}{\overset{\overset{CH_3}{|}}{C}}-Cl$, 2-chloro-2-methylbutane

(2) 1-bromo-16-methylheptadecane

(3)  Cl-⟨O⟩-Cl   (4) 2,4,6-tribromotoluene

(5) $(CH_3)_3CCH_2\underset{\underset{CH_3}{|}}{N}-C_2H_5$, also named N-ethyl-N,2,2-trimethylpropylamine (CA)

(6) N-ethyl-N-isopropyl-tert-butylamine, or 1,1,1'-trimethyltri-ethylamine (CA)

(7) 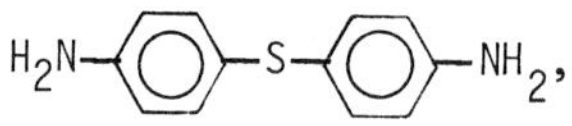 H_2N-⟨O⟩-S-⟨O⟩-NH_2, the amine function has a higher

priority than sulfide. "bis" indicates two identical groups.

(8) bis(m-cyanophenyl) disulfide, or better 3,3'-dithiodibenzo-nitrile

(9) 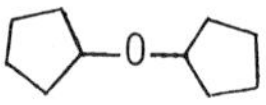 ⟨⟩-O-⟨⟩

(10) isohexyl phenyl ether (CA), 4-methyl-1-phenoxypentane, (isohexyloxy)benzene

(11) $CH_3CH-C(CH_3)_2$
        $|\quad\quad |$
       $NH_2\ SH$

(12) 8-amino-5-phenyl-1-octanol

(13) 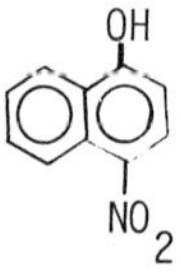 [naphthalene with OH and NO$_2$ substituents]

(14) 7-chloro-2-naphthalenethiol

(15) [structure: benzene ring with H₃C-, -NH₂, and two CH₃ groups]

(16) 2-chloro-6-nitrotoluene

36. 1-bromohexane, 2-bromohexane, 3-bromohexane, 1-bromo-2-methylpentane, 1-bromo-3-methylpentane, 1-bromo-4-methylpentane, 2-bromo-4-methylpentane, 2-bromo-3-methylpentane, 2-bromo-2-methylpentane, 3-bromo-2-methylpentane, 3-bromo-3-methylpentane, 1-bromo-2,2-dimethylbutane, 1-bromo-3,3-dimethylbutane, 1-bromo-2,3-dimethylbutane, 3-(bromomethyl)pentane, 3-bromo-2,2-dimethylbutane, 2-bromo-2,3-dimethylbutane. No unambiguous classical names other than hexyl bromide and isohexyl bromide.

37. (1) $CH_3CCH_2CCH_3$ with two C=O groups

(2) 5-methyl-1,3-cyclohexanedione

(3) $C_6H_5CH_2CH_2C(CH_3)=CHCHO$

(4) 2-methoxy-4,5-dimethylhexanal

(5) $CH_3(CH_2)_4CH(CH_3)-CH(I)CO_2K$

(6) potassium 3-ethyl-2-hydroxyvalerate (CA)

(7) $CH_3CH(CH_3)-CH(CH_3)-CN$

(8) 2-ethylhexanenitrile

(9) $C_6H_5COOCH_2CH_3$

(10) phenyl propionate

(11) [bis(2,6-dichlorobenzoyl) structure with bracket subscript 2, —C(=O)—O—]₂

(12) bis[2-(hexyloxy)-3-methylbutyric] anhydride

(13) [naphthalene with COCl substituent]

(14) 4-methylvaleryl chloride (CA)

(15) [benzene ring with two Cl and CN substituents — 2,6-dichlorobenzonitrile structure]

(16) 2-methyl-1-naphthonitrile

(17) $CH_3CH_2CH_2CH_2CONCH_2CH_3$ with CH₃ on N, also N-ethyl-N-methylvaleramide (CA)

(18) N-phenylcycloheptanecarboxamide; a better name makes use of "anilide" terminology (derived from aniline), cycloheptane-carboxanilide (IUPAC and CA).

Chapter 4

ALKANES: REACTIVE CARBON INTERMEDIATES

1. (1) $CH_3(CH_2)_{14}CH_3$, i.e., $CH_3(CH_2)_7-(CH_2)_7CH_3$

   (2) $(CH_3)_2CHCH_2CH_2CH(CH_3)_2$ (3) 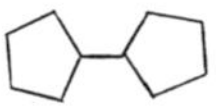 ⬡—⬠   (4) 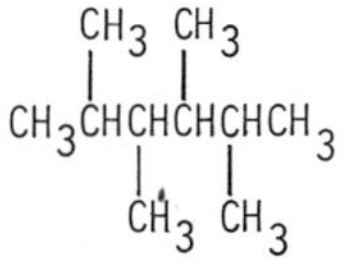 $CH_3\overset{CH_3}{\underset{CH_3}{CH}}\overset{CH_3}{\underset{CH_3}{CH}}CHCH_3$

   Note that coupling takes place at the least substituted carbon atom of the two joined by the double bond.

2. dodecane, decane, and undecane; 2,3-dimethylnonane; 1-cyclopentylhexane

3. (1) ethylbenzene, (2) propylbenzene, (3) heptane

4. $H_2C{=}CH_2 \xrightarrow[\text{heat, pressure}]{H_2,\ Ni} CH_3CH_3$; $H_2C{=}CH_2 \xrightarrow[\text{KOH, AgNO}_3]{B_2H_6}$

   $CH_3CH_2CH_2CH_3$; $HC{=}O \xrightarrow[\text{Zn·Hg—HCl}]{H\ [H]} CH_4$ or $CH_3\underset{\underset{O}{\|}}{C}CH_3 \longrightarrow$

   $CH_3CH_2CH_3$

5. $CH_4 \xrightarrow{O_2} C + H_2O$; $CH_4 \xrightarrow{O_2} CO + H_2O$

6. chloromethane, dichloromethane, trichloromethane, tetrachloromethane

7. 1-chloropentane, 2-chloropentane, 3-chloropentane, 1-chloro-3-methylbutane, 2-chloro-3-methylbutane, 2-chloro-2-methylbutane, 1-chloro-2-methylbutane

8. $2Cl\cdot \longrightarrow Cl_2$, $CH_3\cdot + Cl\cdot \longrightarrow CH_3Cl$, $2CH_3\cdot \longrightarrow CH_3CH_3$

9. Carbon has four valence electrons (−) and four compensating protons (+) in the nucleus; hydrogen is also electrically neutral and thus

we have: H· and ·C̈· giving a neutral free radical:

H:C̈·
 H
 H (with H above C)

On the other hand, if an electron is added to give a carbanion, there is one electron (negative charge) in excess over the positive charges (protons) in the nucleus of carbon:

H:C̈:⊖
 H
 H

If the electron of the free radical is removed to give a carbonium ion, there is one positive charge (proton) in excess in the nucleus, thus, the positive charge on the carbonium ion:

H:C̈⊕
 H
 H

10. The phenyl groups weaken the C—H bond, so that H may more easily be abstracted by a strong base. The anion of the resulting salt is a carbanion.

$(C_6H_5)_3CH + NaNH_2 \longrightarrow (C_6H_5)_3C:^{\ominus} Na^{\oplus} + NH_3$

11. $CH_4 \xrightarrow[\text{heat}]{\text{cat.}} C + 2H_2$

12. $CH_3CH_2CH_3$, $CH_3(CH_2)_2CH_3$, $CH_3(CH_2)_3CH_3$, $CH_3(CH_2)_4CH_3$, $CH_3(CH_2)_5CH_3$, $CH_3(CH_2)_6CH_3$, $CH_3(CH_2)_2CH(CH_3)(CH_2)_2CH_3$, $(CH_3)_3CCH_2CH(CH_3)_2$, $(CH_3CH_2)_2CHCH_2CH_2CH_3$, $CH_3CH_2CH-CHCH_2CH_3$, $(CH_3)_3CCH(CH_3)CH_2CH_3$,
$\qquad\qquad\qquad\qquad\qquad\qquad\qquad\;\;|\quad\;|$
$\qquad\qquad\qquad\qquad\qquad\qquad\qquad CH_3\;CH_3$

[cyclohexane chair], [benzene ring], [benzene ring]—CH_3

13. [cyclohexyl]—$CH(CH_3)_2$ $\xrightarrow[\text{heat}]{\text{Pt cat.}}$ [phenyl]—$CH(CH_3)_2$

14. (1) $C(CH_3)_4$ (2) $(CH_3)_2CH(CH_2)_6\overset{CH_3}{\underset{|}{C}}H-\overset{CH_3}{\underset{|}{C}}HCH_2CH_3$ (note numbering is not 3,4,11 because of the lowest point of first difference)

(3) $CH_3CH_2\overset{CH_3}{\underset{|}{C}}=\overset{}{\underset{|}{C}}CH_2CH_3$
$\qquad\qquad\quad\;\; C_2H_5$

(4) $(CH_3)_2CH(CH_2)_3CH_3$

(5) 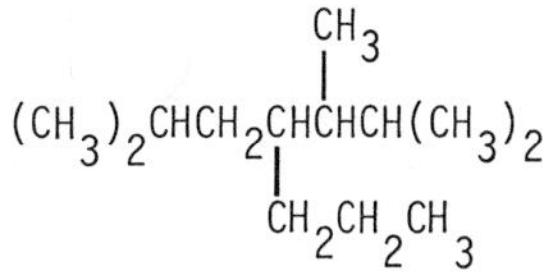 $(CH_3)_2CHCH_2\underset{\underset{CH_2CH_2CH_3}{|}}{\overset{\overset{CH_3}{|}}{C}H}CHCH(CH_3)_2$

(6) 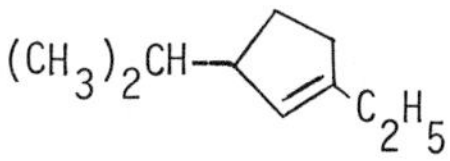 $(CH_3)_2CH$—[cyclobutene ring]—C_2H_5

(7) 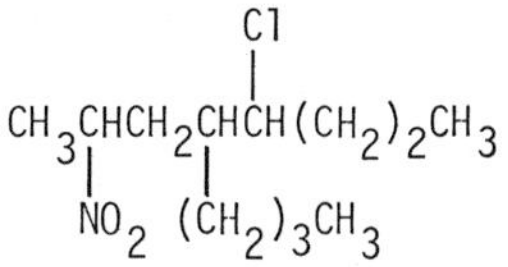 $CH_3\underset{\underset{NO_2}{|}}{C}HCH_2\underset{\underset{(CH_2)_3CH_3}{|}}{C}HCH(CH_2)_2CH_3$ (with Cl on the third carbon)

(8) 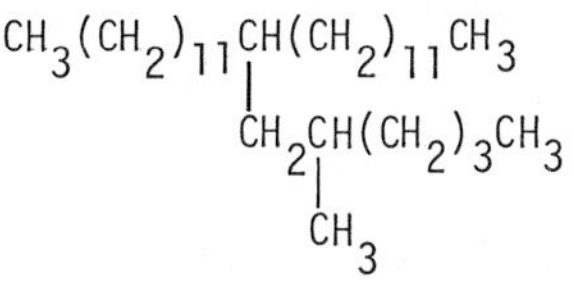 $CH_3(CH_2)_{11}\underset{\underset{\underset{\underset{CH_3}{|}}{CH_2CH(CH_2)_3CH_3}}{|}}{C}H(CH_2)_{11}CH_3$

15. (1) 5-methyl-3-propyl-1-hexene; (2) 2,2-dimethylbutane; (3) 2,2,3,3-tetramethylpentane; (4) 4-vinylcycloheptene; (5) 5-ethyl-2,2,3-trimethyloctane; (6) 1,3-diphenylbutane; (7) 4,5-diethyl-2,7-dimethyloctane; (8) 12-(2,3-dimethylbutyl)triacontane

16. 2-methylpropane, butane, pentane, 2,2-dimethylbutane, 2-methylpentane, hexane

17. (1) 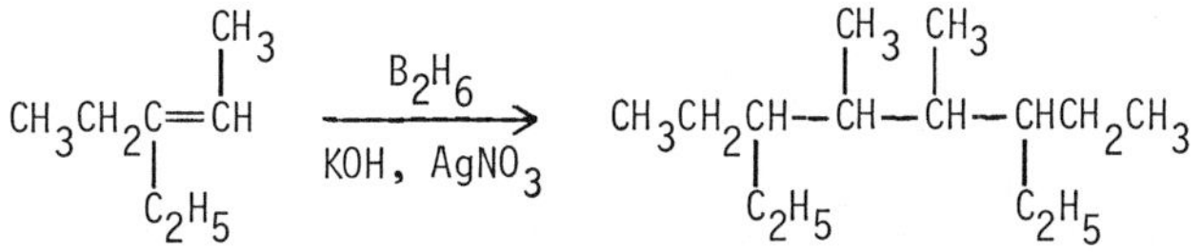 $CH_3CH_2\underset{\underset{C_2H_5}{|}}{C}=CH \xrightarrow[KOH, AgNO_3]{B_2H_6} CH_3CH_2CH\underset{\underset{C_2H_5}{|}}{-}\overset{\overset{CH_3}{|}}{C}H-\overset{\overset{CH_3}{|}}{C}H-\underset{\underset{C_2H_5}{|}}{C}HCH_2CH_3$

(2) Products are 1-nitropropane, 2-nitropropane, nitroethane, and nitromethane.

(3) tert-Butyl bromide is the major product over isobutyl bromide. This is so because the order of stability of free radicals is: tertiary > secondary > primary.

(4) $CO_2 + H_2O$ (5) hexane   (6) 2,3-dimethylbutane

(7)  ◇—◇ (bicyclobutyl)   (8) 1,3-diphenylpropane

18. (1) A readily oxidized substance which thus can accomplish the reduction of the other reactant (substrate). Hydrogen in catalytic hydrogenation.   (2) Such a mechanism involves an atom or a free radical as an initiator. This abstracts an atom from the organic reactant and sets up a process that is self-perpetuating. An example is scheme 4.8.   (3) An exothermic reaction is one in which heat is evolved. The potential energy of the products is less than the reactants. See energy profile in Fig. 4.3; either the overall reaction or step 3.   (4) A highly reactive carbon intermediate where the carbon atom bears a formal positive charge, $(CH_3)_3C^{\oplus}$; the order of stability of carbonium ions is important in postulation of reaction mechanisms: tertiary > secondary > primary. (5) The heat of reaction ΔH is the sum of the energy content of the products (sum of bond dissociation energies) and the reactants; see Fig. 4.3 for ΔH react.   (6) A transition state is a non-isoable intermediate where bonds are being made and broken simultaneously:

$$\overset{\delta\cdot}{H_3C}\cdots H\cdots \overset{\delta\cdot}{Br}$$

as represented by the peak of the energy profile (I) for step 2 in Fig. 4.3.   (7) The energy of activation is the amount of energy required to convert the reactants to the transition state. See E_{act} (2,3) in Fig. 4.3   (8) A carbanion is a sometimes highly reactive trivalent species of carbon with a nonbonding pair of electrons and bearing a negative charge:

$$(C_6H_5)_3C:^{\ominus}$$

(9) A catalyst is a substance which increases the rate of a reaction substantially. It is required in only small amount and is itself not changed in the course of the reaction.   (11) An oxidizing agent is a substance which is easily reduced and thus can accomplish the oxidation of the other reactant. Oxygen in the combustion of hydrocarbons.   (12) An endothermic reaction is one which absorbs heat and where the potential energy of the products is greater than the reactants: See Fig. 4.3 for step 2.

(10) A carbon free radical is a highly reactive species containing seven electrons about C, the unpaired H_2, H_3C

19. (1) 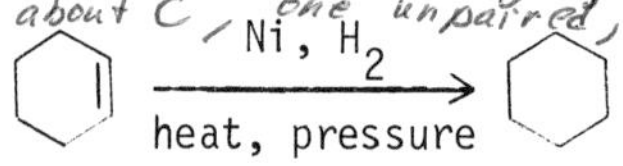 ⌬ —Ni, heat, pressure→ ⌬

(2) 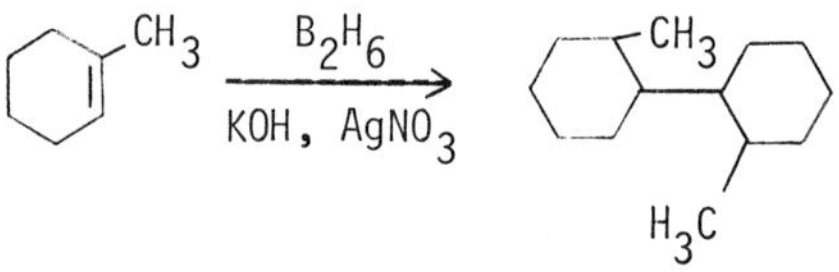 ⌬-CH₃ —B₂H₆, KOH, AgNO₃→ ⌬-CH₃ with H₃C substituent

(3) Either reactions (1) or (2)

(4) $CH_3\underset{CH_3}{\overset{CH_3}{\underset{|}{\overset{|}{C}}}}CH_3$ —Cl_2, light→ $(CH_3)_3CCH_2Cl$

(5) $2CH_3CH_3$ —$7O_2$→ $4CO_2 + 6H_2O$; these reactions can be easily balanced.

(6) $(CH_3)_2CH-\underset{O}{\overset{||}{C}}-\underset{CH_3}{\overset{|}{C}}HCH_2CH_3$ —[H], Zn·Hg—HCl→ $(CH_3)_2CHCH_2\underset{CH_3}{\overset{|}{C}}HCH_2CH_3$

20.
$Cl_2 \overset{h\nu}{\longrightarrow} 2Cl\cdot$; $Cl\cdot + C_6H_5CH_3 \longrightarrow C_6H_5CH_2\cdot + HCl$;

$C_6H_5CH_2\cdot + Cl_2 \longrightarrow C_6H_5CH_2Cl + Cl\cdot$

21. $Br_2 \longrightarrow 2Br\cdot$

$Br\cdot + CH_4 \longrightarrow CH_3Br + H\cdot$

104 kcal   70   + 34

$$H\cdot + Br_2 \longrightarrow HBr + Br\cdot$$
$$46 \qquad 87 \qquad -41$$

Since this alternative mechanism is more endothermic by 17 kcal (34 - 17) in step 2 over that given in 4.10, it would not be favored.

22. $$C_6H_5CH_2-H + Br_2 \longrightarrow C_6H_5CH_2Br + HBr$$
$$85 + 46 = +131, \quad -70 \quad -87 = -157$$

$$\Delta H = (-157 + 131) = -26 \text{ kcal}$$

The CH_3- group of toluene would be easier to brominate than methane, since the reaction would be more exothermic; -26 kcal versus -7 kcal.

ALKENES AND ALKYNES: ADDITION POLYMERIZATION

1. (1) $CH_3CH_2CH\!=\!CH_2$, 1-butene (2) $CH_3C\!\equiv\!CH$, 1-propyne

   (3) $CH_3CH\!=\!CH_2$, 1-propene (4) $CH_3C\!\equiv\!CCH_3$, 2-butyne

   (5) $(CH_3)_2C\!=\!CH_2$, 2-methyl-1-propene

   Derived classical names for acetylenes are obtained by identifying substituent groups replacing the hydrogen atoms of parent acetylene. In IUPAC names, such a 1-propyne and 1-propene, where the functional group can only be one position, the number is unnecessary.

2. (1) $\overset{2}{(CH_3)_2C}\!=\!CH(CH_2)_2\overset{6}{\underset{\underset{CH_3}{|}}{C}}\!=\!CH(CH_2)_2\overset{10}{\underset{\underset{CH_3}{|}}{C}}\!=\!CH(CH_2)_2\overset{14\ 15}{CH}\!=\!\overset{18}{\underset{\underset{CH_3}{|}}{C}}(CH_2)_2CH\!=\!$
   $\overset{19}{\underset{\underset{CH_3}{|}}{C}}(CH_2)_2\overset{22\ 23}{CH}\!=\!C(CH_3)_2$, $C_{30}H_{50}$

   (2) squalene, 2,6,10,15,19,23-hexamethyl-2,6,10,14,18,22-tetracosahexaene; bombykol, 10,12-hexadecadien-1-ol. Accurately, the name bombykol is reserved for 10 $\underline{E}$, 12 $\underline{Z}$ isomer (10 $\underline{trans}$, 12 $\underline{cis}$)(Sections 6.2, 6.7).

3. (1) $CH_3\underset{\underset{Br}{|}}{CH}\!-\!\underset{\underset{Br}{|}}{CH}CH(CH_3)_2 \xrightarrow{Zn} CH_3CH\!=\!CHCH(CH_3)_2 + ZnBr_2$

   (2) $CH_3CH_2\underset{\underset{Br}{|}}{CH}\!-\!\underset{\underset{Br}{|}}{CH}(CH_2)_3CH_3 \xrightarrow{Zn} CH_3CH_2CH\!=\!CH(CH_2)_3CH_3 + ZnBr_2$

4. $BrCH_2CH_2\overset{\oplus}{}\overset{:\ddot{O}H_2}{\curvearrowleft} \longrightarrow BrCH_2CH_2\!-\!\underset{\underset{H}{|}}{\overset{\oplus}{O}}H \longrightarrow BrCH_2CH_2OH + \overset{\oplus}{H}$

$$CH_3(CH_2)_3CH{=}CH_2 \xrightarrow{Br_2} CH_3(CH_2)_3\overset{\oplus}{C}HCH_2Br \xrightarrow{HOCH_3} CH_3(CH_2)_3CHCH_2Br$$
$$\downarrow Cl^{\ominus} \qquad\qquad\qquad\qquad |$$
$$CH_3(CH_2)_3CHClCH_2Br \qquad\qquad OCH_3$$

1-bromo-2-methoxyhexane or 1-(bromo-methyl)pentyl methyl ether (CA)

5.

$$CH_3CH{=}CH_2 \xrightarrow[\text{usual temp.}]{Cl_2} CH_3CHCH_2Cl$$
$$\qquad\qquad\qquad\qquad |$$
$$\qquad\qquad\qquad\qquad Cl$$

$$CH_3CH{=}CH_2 \xrightarrow[500°]{Cl_2} ClCH_2CH{=}CH_2 + HCl$$

3-chloro-1-propene, allyl chloride

Note that it is a methyl or methylene group directly adjacent to the double bond which is reactive enough to undergo substitution; $CH_3CHClCH{=}CH_2$ would be formed in preference to $ClCH_2CH_2CH{=}CH_2$.

6. $F_3CCH_2CH_2Br$

The F groups have an inductive effect directly opposite to that of CH_3- groups, electron-attracting rather than electron-donating. This creates a partial positive charge on the carbon to which it is attached. Markownikoff addition would place a positive charge on the adjacent carbon, a very unstable situation. Consequently, anti-Markownikoff addition would occur. This illustrates why it is important that Markownikoff's rule be modified in terms of modern organic theory.

7.

$$\begin{array}{c} CH_3 \\ | \\ CH_3C{=}CHCH_3 \end{array} \xrightarrow{HBr} \begin{array}{l} \longrightarrow (CH_3)_2\underset{Br}{\overset{|}{C}}{-}CH_2CH_3 \\ \\ \underset{\text{peroxide}}{\longrightarrow} (CH_3)_2CH{-}\underset{Br}{\overset{|}{C}}HCH_3 \end{array}$$

8. $6CH_4 + O_2 \xrightarrow{\text{heat}} 2HC{\equiv}CH + 2CO + 10H_2$

9.

$$\begin{array}{c} CH_3 \\ | \\ CH_3C{-}Br \\ | \\ H_2C \\ \quad H \quad :C{\equiv}CH \\ \qquad \overset{\ominus}{\underset{Na^{\oplus}}{}} \end{array} \longrightarrow \begin{array}{c} CH_3 \\ | \\ CH_3C \\ \parallel \\ CH_2 \end{array} + HC{\equiv}CH{\uparrow} + NaBr$$

10. $(CH_3)_2CHCH=CH_2 \xrightarrow{Br_2} (CH_3)_2CHCHBr-CH_2Br \xrightarrow{NaNH_2}$

   $(CH_3)_2CHC\equiv CH$

   (2)   (1)

11. $CH_3CH_2CH_2C\equiv CH \xrightarrow{Na} CH_3CH_2CH_2C\equiv CNa + H_2\uparrow$ Or the reaction with $Ag^{\oplus}$ to give an insoluble precipitate of silver acetylide.

   $CH_3CH_2C\equiv CCH_3 \xrightarrow{Na}\!\!\!/\!\!\!/\!\!\!\to$ NO REACTION

12. $CH_2=CHCl \xrightarrow{HCl} CH_3CHCl_2$

   (1) $CH_3C\equiv CH \xrightarrow{Cl_2} CH_3CCl=CHCl \xrightarrow{Cl_2} CH_3CCl_2CHCl_2$

   (2) $CH_3C\equiv CH \xrightarrow{HBr} CH_3CBr=CH_2 \xrightarrow{HBr} CH_3CBr_2CH_3$

13. Head-to-head and tail-to-tail orientations involve less stable primary free radicals.

   $\sim\!\!\sim\!\!\sim CH_2\underset{Cl}{CH}\cdot \;\; \underset{Cl}{CH}=CH_2 \longrightarrow \sim\!\!\sim\!\!\sim CH_2\underset{Cl}{CH}-\underset{Cl}{CH}CH_2\cdot \longrightarrow$

                                       (1)

                              Tail-to-tail involves a primary free radical.

14. $CH_2=\underset{CN}{CH} \xrightarrow{RO\cdot} ROCH_2\underset{CN}{CH}\cdot \xrightarrow{CH_2=CHCN} ROCH_2\underset{CN}{CH}CH_2\underset{CN}{CH}\cdot$, etc.

   $ROCH_2\underset{CN}{CH}\!-\!\!\left[\!-CH_2\underset{CN}{CH}\!-\!\right]_n\!\!-\!CH=CHCN$

15. (1) $H_2C=CH-CH=CH-CH=CH_2$ (2) [cycloheptatriene ring]   (3) [cyclohexadiene ring]

   (4) * $H_2C=CH-CH_2CH=CH_2$ (5) * [cyclohexadiene ring]

   Starred formulas are nonconjugated.

16. * $ClCH_2CH=CH-CH=CHCH_2Cl$, * $ClCH_2CH=CHCHCH=CH_2$,
                                                    |
                                                    Cl

$ClCH_2CHCH=CH-CH=CH_2$, $H_2C=CH-CH-CH-CH=CH_2$
        | | |
        Cl   Cl Cl

Starred formulas involve conjugate addition.

17. [cyclobutadiene] + $\overset{H}{\underset{C-C_6H_5}{\overset{|||}{C}}}$ → [bicyclic structure with phenyl] $\xrightarrow{heat}$ [biphenyl]

                                                unstable   biphenyl stable

The unstable cycloaddition product formed initially is like that proposed for benzene in 1867 by J. Dewar. Unlike benzene it is a very unstable structure.

13. (Continued)

$\sim\sim\sim CHCH_2\cdot + CH_2=CH \longrightarrow \sim\sim CHCH_2CH_2CH\cdot$
       | | | |
       Cl   Cl   Cl   Cl

    (2)

Head-to-head requires (2), a primary free radical, as a precursor.

18. $\left[\begin{array}{c}Cl\diagdown\diagup CH_2\\ C=C\\ \sim CH_2\diagup\diagdown H\end{array}\right.$ $\left.\begin{array}{c}-CH_2\diagdown\diagup H\\ C=C\\ Cl\diagup\diagdown CH_2-CH_2\end{array}\right.$ $\left.\begin{array}{c}Cl\diagdown\diagup CH_2\\ C=C\\ \diagup\diagdown H\end{array}\right.$ $\left.\begin{array}{c}-CH_2\diagdown\diagup H\\ C=C\\ Cl\diagup\diagdown CH_2\sim\end{array}\right]_n$

19. $CH_2=\underset{\underset{CH_3}{|}}{\overset{\overset{CH_3}{|}}{C}}$ → $\left[-CH_2-\underset{\underset{CH_3}{|}}{\overset{\overset{CH_3}{|}}{C}}-\right]_n$ Butyl rubber is a vinyl polymer.

20. (1) $CH_3CH_2CH_2CH=CHCH_3$

    (2) $CH_3CH_2CH_2\underset{\underset{C_3H_7}{|}}{C}=\underset{\underset{C_3H_7}{|}}{C}CH_2CH_2CH_3$, 4,5-dipropyl-4-octene

(3) $CH_3CH_2C{\equiv}CCH(CH_3)_2$, 2-methyl-3-hexyne

(4) $CH_3(CH_2)_4C{\equiv}C(CH_2)_4CH_3$

(5) $CH_3(CH_2)_{15}\underset{\underset{C_6H_5}{|}}{C}{=}\underset{\underset{C_2H_5}{|}}{C}CH_2CH_3$

(6) $CH_3(CH_2)_2C{\equiv}CCH_2CH_3$

(7) $CH_2{=}\underset{\underset{Cl}{|}}{C}{-}\underset{\underset{C_6H_5}{|}}{C}HCH_3$

(8) $CH_3CH_2\underset{\underset{C_6H_5CH_2}{|}}{C}HC{\equiv}CBr$

21. (1) 3-chloro-1-butene; C=C takes precedence over chloro substituent in numbering; (2) 5-bromo-1-pentene; (3) 3-propyl-1-heptene, not 4-vinyloctane; (4) 3,5-diethyl-3-heptene; (5) 3-ethyl-2-methyl-2-hexene; the longest chain proceeds through the propyl group for six carbon atoms, rather than straight across for five carbon atoms; (6) 4-butyl-3-ethyl-3-nonene

22.

(1) $CH_3CH_2OH \xrightarrow[\text{heat}]{Al_2O_3} H_2C{=}CH_2$

(2) $CH_3CH_2\underset{\underset{OH}{|}}{C}HCH_3 \xrightarrow[\text{heat}]{Al_2O_3} CH_3CH{=}CHCH_3$

(3) $CH_3CH_2Cl \xrightarrow[\text{EtOH}]{KOH} H_2C{=}CH_2$

(4) $H_2C{=}CH_2 + H_2O \xrightarrow[KMnO_4]{[O]} HOCH_2CH_2OH$

(5) $HC{\equiv}CH + Ag^{\oplus} \longrightarrow HC{\equiv}CAg\downarrow$

23.

(1) $CH_3(CH_2)_5CH{=}CHCH_3 \xrightarrow[\text{heat, pressure}]{H_2\text{-Pt}} CH_3(CH_2)_7CH_3$

(2) $CH_3CH_2CH{=}CH_2 \xrightarrow[Br^{\ominus}]{Cl_2} CH_3CH_2\underset{\underset{Br}{|}}{C}HCH_2Cl + CH_3CH_2\underset{\underset{Cl}{|}}{C}HCH_2Cl$

(3) $(CH_3)_2C{=}CH_2 \xrightarrow{HCl} (CH_3)_3CCl$

(4) $CH_3\underset{\underset{CH_3}{|}}{C}{=}CHCH_2CH_3 \xrightarrow[(CH_3)_3C-O-O-C(CH_3)_3]{HBr} (CH_3)_2CHCH\underset{\underset{Br}{|}}{C}H_2CH_3$

(5) $(CH_3)_2CHCH_2CH_2CH_2OH \xrightarrow[\text{heat}]{Al_2O_3} (CH_3)_2CHCH_2CH{=}CH_2$

(6) $CH_3CH_2\underset{\underset{Br}{|}}{\overset{\overset{CH_3}{|}}{C}}CH_3 \xrightarrow[EtOH]{KOH} CH_3CH=C(CH_3)_2$

(7) $CH_3C\equiv CH \xrightarrow{HI} CH_3\underset{\underset{I}{|}}{C}=CH_2 \longrightarrow (CH_3)_2CI_2$

(8) $CH_3CH_2C\equiv CH \xrightarrow[HgSO_4]{CH_3CO_2H} CH_3CH_2\underset{\underset{OCOCH_3}{|}}{C}=CH_2$

(9) $CH_3CH_2CH=CHCH_3 \xrightarrow[[O], H_2O]{KMnO_4} CH_3CH_2\underset{\underset{OH}{|}}{CH}-\underset{\underset{OH}{|}}{CH}CH_3$

(10) $(CH_3)_2CHCH_2CH_2CH_2Br \xrightarrow{NaC\equiv CH} (CH_3)_2CHCH_2CH_2CH_2C\equiv CH$

24.

(1) $n\ CH_2=CHOCOCH_3 \xrightarrow[heat]{peroxide} \left[-CH_2\underset{\underset{OCOCH_3}{|}}{CH}-\right]_n$

(2) $n\ CH_2=CHC_6H_5 \xrightarrow[heat]{peroxide} \left[-CH_2\underset{\underset{C_6H_5}{|}}{CH}-\right]_n \quad n = \sim 5000$

(3) $n\ 150\ H_2C=\underset{\underset{C_6H_5}{|}}{CH} + n\ H_2C=\underset{\underset{\underset{CH_3}{|}}{C=O}}{CH} \longrightarrow \left[-(H_2C-\underset{\underset{C_6H_5}{|}}{CH})_{150}CH_2-\underset{\underset{\underset{CH_3}{|}}{C=O}}{CH}-\right]$

One ketone group for each 150 styrene groups gives just the right degree of oxidative vulnerability for a biodegradable polymer.

25.

(1) a. $\overset{nucleophilic}{CH_3CH_2^{\delta-}MgBr} + \overset{electrophilic}{HOH^{\delta+}} \longrightarrow CH_3CH_3 + MgBr(OH)$

b. $CH_3CH_2MgBr + HC\equiv CH \longrightarrow CH_3CH_3 + BrMgC\equiv CH$

(2) a. $CH_3CHBrCH_2Br$ b. $CH_3CBr_2CHBr_2$

(3) a. $CH_3C\equiv CCH_3$ b. $H_2C=\underset{\underset{CH_3}{|}}{C}-\underset{\underset{CH_3}{|}}{C}=CH_2$

26.

(1) $(CH_3)_2CHBr \xleftarrow{HBr} CH_3CH=CH_2 \xleftarrow[EtOH]{KOH} CH_3CH_2CH_2Br$

$CH_3CH_2CH_2Br \xleftarrow[\text{peroxide}]{HBr} CH_3CH=CH_2 \xleftarrow[EtOH]{KOH} CH_3\overset{Br}{\underset{|}{C}}HCH_3$

(3) $CH_3CH_2\underset{\underset{OH}{|}}{C}HCH_2OH \xleftarrow[KMnO_4]{[O], H_2O} CH_3CH_2CH=CH_2 \xleftarrow[(2)\ KOH-EtOH]{(1)\ HBr} CH_3CH_2CH_2CH_2OH$

(4) $CH_3CH_2\underset{\underset{CH_3}{|}}{\overset{\overset{CH_3}{|}}{C}}-Cl \xleftarrow{HCl} CH_3CH_2\overset{\overset{CH_3}{|}}{C}=CH_2 \xleftarrow[EtOH]{KOH} CH_3CH_2\overset{\overset{CH_3}{|}}{C}HCH_2Cl$

(5) $CH_2=CCl_2 \xleftarrow{Ca(OH)_2} ClCH_2CHCl_2 \xleftarrow{Cl_2} CH_2=CHCl$

The product is vinylidene chloride, one of the monomers (along with vinyl chloride) for Saran (see Pb. 5.27).

(6) $C_2H_5C{\equiv}CC_2H_5 \xleftarrow[\text{2 times}]{Na,\ C_2H_5Br} HC{\equiv}CH$

(7) $(CH_3)_3CCH_2C{\equiv}CH \xleftarrow[EtOH]{KOH} (CH_3)_3CCH_2\underset{\underset{Br}{|}}{C}HCH_2Br \xleftarrow{Br_2} (CH_3)_3CCH_2CH=CH_2$

(8) $(CH_3)_2\underset{\underset{OH}{|}}{C}CH_2Cl \xleftarrow{Cl_2,\ H_2O} (CH_3)_2C=CH_2 \xleftarrow[EtOH]{KOH} (CH_3)_2CHCH_2Cl$

Under proper conditions the chloro alcohol can be made the major product.

27. $CH_2=CHCl + CH_2=CCl_2 \longrightarrow \left[-CH_2CH-CH_2\underset{\underset{}{}}{C}-\right]_n$ with Cl and Cl_2 substituents

28. $2\ HC{\equiv}CH \xrightarrow[\text{heat}]{Cu_2Cl_2,\ KCl,\ H^{\oplus}} H_2C=CH-C{\equiv}CH$ vinylacetylene
1-buten-3-yne

Double bond takes precedence over triple bond in IUPAC name if a lower number can be maintained.

$4\ HC{\equiv}CH \xrightarrow[\text{heat, pressure}]{Ni(CN)_2}$ cyclooctatetraene

29. In the first step with the nonconjugated system, only 1,2-addition is possible, whereas with the isomeric conjugated system 1,4-addition is the primary reaction.

$$H_2C{=}CH(CH_2)_3\overset{2}{CH}{=}\overset{1}{CH_2} \xrightarrow{Br_2} H_2C{=}CH(CH_2)_3\underset{Br}{\overset{|}{CH}}CH_2Br$$

$$CH_3CH_2CH_2\overset{4}{CH}{=}CH{-}\overset{1}{CH}{=}CH_2 \xrightarrow{Br_2} CH_3CH_2CH_2\underset{Br}{\overset{|}{CH}}CH{=}CHCH_2Br$$

chapter 6

STEREOCHEMISTRY: GEOMETRIC AND OPTICAL ISOMERISM

1. $CH_3(CH_2)_4CH_3$, $CH_3CH_2CH(CH_3)CH_2CH_3$, $(CH_3)_2CHCH_2CH_2CH_3$,
    hexane   3-methylpentane   2-methylpentane

    $(CH_3)_3CCH_2CH_3$ $(CH_3)_2CHCH(CH_3)_2$
    2,2-dimethylbutane   2,3-dimethylbutane

2. (1) 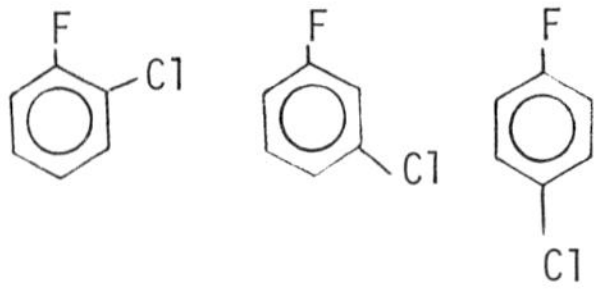 [1-chloro-2-fluorobenzene, 1-chloro-3-fluorobenzene structures with F and Cl substituents]

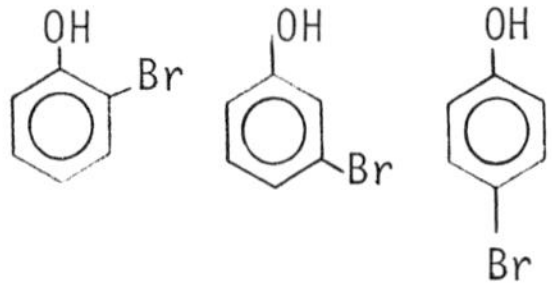

    1-chloro-2-fluorobenzene
    1-chloro-3-fluorobenzene, etc.

    (2) [o-, m-, p-bromophenol structures with OH and Br substituents]

    o, m, and p-bromophenol

3. 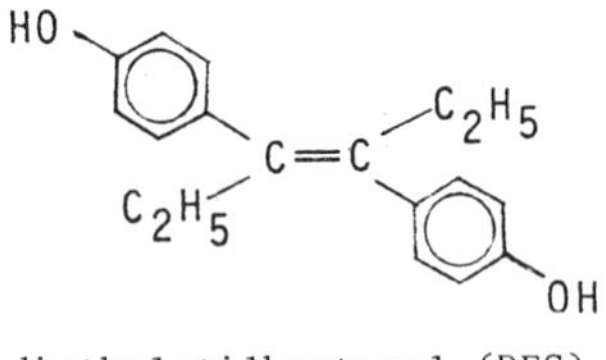 [diethylstilbesterol structures — cis and trans isomers with C=C, C_2H_5 groups and hydroxyphenyl rings]

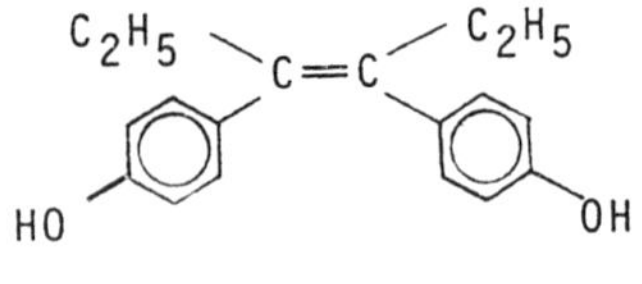

    diethylstilbesterol (DES)

4. (3) $CH_3CH_2\overset{*}{C}HCH_3$ $CH_3CH_2CH_2\overset{*}{C}HCH_3$
              | |
              OH   Cl

    (3) $C_6H_5\overset{*}{C}H-\overset{*}{C}HCO_2H$ (4) $HOCH_2\overset{*}{C}H-\overset{*}{C}H-\overset{*}{C}HCHO$
                | | | | |
                OH   OH   OH OH OH

    Formulas should be written out for clarity.

5.
```
     CO₂H            CO₂H
      |               |
   H—C—H           H—C—H
      |               |
     C₆H₅            C₆H₅
     (1)             (2)
```

(1) and (2), identical formulas, are mirror images. Clearly, symmetrical molecules can also have mirror images; however, because of the symmetry the mirror images would be superimposable and would not show optical isomerism.

6.

$([\alpha]_D^{20} \ -50.6°)$

$[\alpha]_D^{20}\ +50.6°$ $[\alpha]_D^{19}\ -50.3°$

```
        CH₃                        CH₃
         |                          |
        / \                        / \
    H /C₆H₅\ Cl             Cl /C₆H₅\ H
       (+)                        (-)
```

bp 85° (20mm)   bp 85° (20mm)

d_4^{20} 1.0631 (1.0632) d_4^{20} 1.0632

(1a)   (1b)

(1) (±) $[\alpha]_D^{20}\ -0°-$

bp 81-82° (17mm); d_4^{20} 1.0620

[bp 85° (20mm); d_4^{20} 1.0632]

7. If one were not informed of the absolute configuration, one would have a 50-50 chance of guessing it. As we outlined the answer to Pb. 6.6, we guessed wrong. Bijvoet's work and chemical interrelationships establish the (-)-isomer as being of the D-configuration. We shall redraw the formulas as Fischer projection formulas.

```
      CH₃                      CH₃
       |                        |
   H—C—Cl                  Cl—C—H
       |                        |
      C₆H₅                     C₆H₅
  [α]_D^{20}  -50.6°          +50.6°
```

8. CH_3CH_2, OH, $ClCH_2$, $CH(CH_3)_2$, CH_2CH_3,

    $ClCH_2\underset{\underset{CH_3}{|}}{\overset{\overset{CH_3CH_2}{|}}{C}}-\underset{\underset{}{|}}{\overset{\overset{}{}}{C}}HCH_3$

    with OH on the second carbon

    Newman projection: front carbon with CH_2Cl (up), CH_3CH_2 (left), HO (lower-left); back carbon with $CH(CH_3)_2$ (right)

9. | D-threose | L-threose | D-erythrose | L-erythrose |
    |---|---|---|---|
    | CHO | CHO | CHO | CHO |
    | HO—C—H | H—C—OH | H—C—OH | HO—C—H |
    | H—C—OH | HO—C—H | H—C—OH | HO—C—H |
    | CH_2OH | CH_2OH | CH_2OH | CH_2OH |

10.

    $\underset{C_6H_5}{H}C=C\underset{CO_2H}{H}$ mp 66-68° $\xrightarrow{Br_2}$

    $\begin{array}{c} C_6H_5 \\ Br-C-H \\ H-C-Br \\ CO_2H \end{array}$ $\begin{array}{c} C_6H_5 \\ H-C-Br \\ Br-C-H \\ CO_2H \end{array}$

    mp 93.5-95°

    cis-cinnamic acid
    Z-cinnamic acid

    $\underset{H}{C_6H_5}C=C\underset{CO_2H}{H}$ mp 133° trans or E-cinnamic acid $\xrightarrow{Br_2}$

    $\begin{array}{c} C_6H_5 \\ H-C-Br \\ H-C-Br \\ CO_2H \end{array}$ $\begin{array}{c} C_6H_5 \\ Br-C-H \\ Br-C-H \\ CO_2H \end{array}$

    mp 203-204°

    The melting points of the individual optical isomers apparently have not been reported. There is no basis for assigning the optical rotations to isomers (absolute configuration) from the data given.

11. Newman projection (1): front — HO (upper-left), H (up), Cl (lower-left), C_6H_5 (down); back — H, C_6H_3, C_6H_5

     Newman projection (2): front — Cl (up), HO (upper-left), C_6H_5 (lower-left); back — H, H, C_6H_5

     Conformations (1) and (2) would be less stable because the phenyl groups are closer together. Conformation (2) would be less stable than (1) because in (2), the forward C_6H_5 is opposed by two larger groups, C_6H_5 and OH, than in the case of (1), C_6H_5 and H.

12. 3, 8, 4; 4, 16, 8; 5, 32, 16; 6, 64, 32; 7, 128, 64; 8, 256, 128; 9, 512, 256; 10, 1024, 512

13.
$$\overset{5}{\underset{*}{HOCH_2}}\overset{4}{\underset{*}{CH}}-\overset{3}{\underset{*}{CH}}-\overset{2}{CH}\overset{1}{C}=\overset{H}{O}$$
with OH, OH, OH on carbons 4, 3, 2

C_1, -CHO: nonasymmetric C atom, two similar points of attachment in C=O

*C_2 dissimilar asymmetric C atom: H-, -OH, -CHO, $HOCH_2CH-CH-$ with OH, OH

*C_3 dissimilar asymmetric C atom: H-, -OH, $HOCH_2CH-$ (OH), -CHCHO (OH)

*C_4 dissimilar asymmetric C atom: H-, -OH, $HOCH_2-$, -CHCHCHO (OH OH)

C_5 $HOCH_2-$: nonasymmetric C atom, two similar points of attachment to H.

14. enantiomer of (+)-B·(+)-AH and of (-)-B·(+)-AH

(±)-B + (-)-AH ⟶ (-)-B·(-)-AH + (+)-B·(-)-AH

15.
(1) $NH_4^{\oplus}Cl^{\ominus}$ (2) $NH_4^{\oplus \ominus}O_2CCH_3$ (3) $(\pm)\text{-}C_6H_5\overset{CH_3}{\underset{|}{CH}}-NH_3^{\oplus \ominus}O_2CCH_3$

(4) $(+)\text{-}C_6H_5\overset{CH_3}{\underset{|}{CH}}-NH_3^{\oplus \ominus}O_2C\text{-}T\text{-}(+) + (-)\text{-}C_6H_5\overset{CH_3}{\underset{|}{CH}}-NH_3^{\oplus \ominus}O_2-T\text{-}(+) \downarrow$

$T = -\overset{OH}{\underset{H}{C}}-\overset{H}{\underset{OH}{C}}-CO_2H$

(5) $NH_3\uparrow + NaCl + H_2O$

(6) $NH_3\uparrow + CH_3CO_2Na + H_2O$

(7) $(+)\text{-}C_6H_5\overset{CH_3}{\underset{|}{CH}}-NH_2 + (+)\text{-}T-CO_2Na + H_2O;$

$(-)\text{-}C_6H_5\overset{CH_3}{\underset{|}{CH}}-NH_2 + (+)\text{-}T-CO_2Na + H_2O$

16. (1) Stereoisomers are isomers resulting from different spatial arrangements of atoms or groups:

$$\begin{array}{cc} \underset{H}{\overset{CH_3CH_2}{>}}C=C\underset{H}{\overset{CH_2CH_3}{<}} & \underset{H}{\overset{CH_3CH_2}{>}}C=C\underset{CH_2CH_3}{\overset{H}{<}} \\ \underline{\text{cis-3-hexene}} & \underline{\text{trans-3-hexene}} \\ \text{or } Z & \text{or } E \end{array}$$

Stereoisomerism embraces both geometric isomerism and optical isomerism.

(2) A <u>trans</u>-isomer is one of a pair of geometric isomers, in which a pair of like groups is situated on opposite sides of a double bond—see (1).

(3) A racemic modification is composed of equal parts of its constituent enantiomers and is thus optically inactive. It is usually a molecular compound rather than a mixture:

$$\begin{array}{ccc} \overset{CO_2H}{\underset{CH_3}{\overset{|}{H-C-OH}}} & \cdot & \overset{CO_2H}{\underset{CH_3}{\overset{|}{HO-C-H}}} \\ & [\alpha]_D = 0 & \\ (-) & 50\text{-}50 & (+) \end{array}$$

(4) The term "dextrorotatory" refers to an optical isomer which rotates plane-polarized light to the right. This is most frequently indicated by a (+) sign; (+)-lactic acid in (3) above is an example.

(5) <u>threo</u>-Configuration involves placing a like atom or group on opposite sides of a Fischer projection formula of a substance involving two dissimilar asymmetric C atoms:

$$\begin{array}{cc} \overset{C_6H_5}{\underset{C_6H_5}{\overset{|}{H-C-NH_2}}} & \overset{C_6H_5}{\underset{C_6H_5}{\overset{|}{H-C-NH_2}}} \\ \underline{\text{threo}} & \underline{\text{erythro}} \end{array}$$

(6) Diastereomers are optical isomers which are not mirror images of one another. For example, the <u>erythro</u> and <u>threo</u> isomers in (5) are diastereomers.

(7) Molecular asymmetry involves an asymmetric molecule containing no asymmetric carbon atom. Asymmetrically substituted allenes is an example (6.20).

(8) A <u>meso</u> isomer is a symmetrical molecule; it is an optical isomer rendered optically inactive by internal compensation. <u>meso</u>-Tartaric acid is an example:

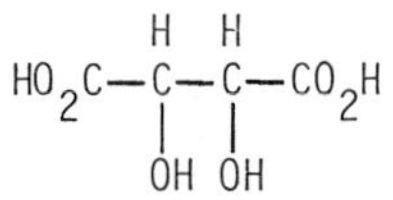

$$HO_2C-\underset{OH}{\underset{|}{C}}-\underset{OH}{\underset{|}{C}}-CO_2H$$
$$HH$$

(9) Absolute configuration involves the assignment of a configurational structure to a stereoisomer. Most absolute configurations are known by their relationship to (+)-tartaric acid.

17. 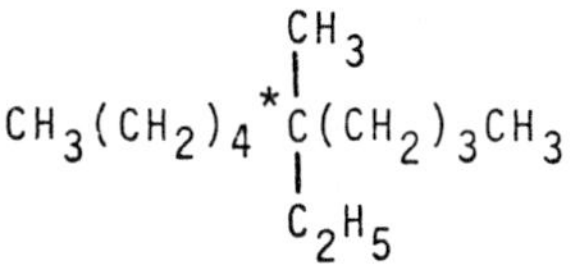 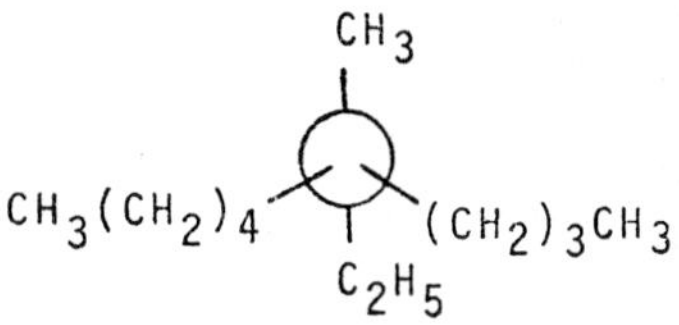 $CH_3(CH_2)_4\overset{*}{\underset{C_2H_5}{\underset{|}{C}}}\overset{CH_3}{\overset{|}{}}(CH_2)_3CH_3$

Newman/wedge projection: $CH_3(CH_2)_4$ — C(CH$_3$)(C$_2$H$_5$) — $(CH_2)_3CH_3$

$CH_3(CH_2)_3$ — C(CH$_3$)(C$_2$H$_5$) — $(CH_2)_4CH_3$

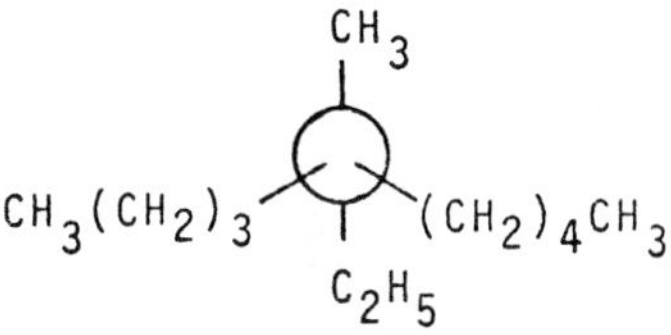

Alkanes of this type have been shown to have a very small optical rotation. This is because of the very similar nature of the four different groups.

18. (1) NH_2, CO_2H, CH_3, H (2) NH_2, CO_2H, CH_2OH, H
    (3) Cl, OCH_3, CH_2CH_3, CH_3 (4) CO_2H, C_6H_5, CH_3, H

19. (1) 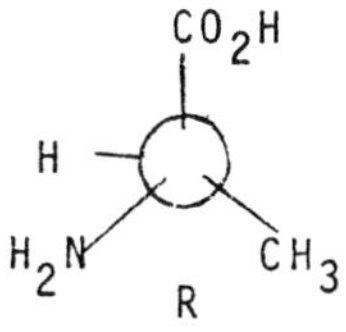 CO_2H / H / H_2N / CH_3 — R
    (2) CO_2H / H / $HOCH_2$ / NH_2 — S
    (3) OCH_3 / H_3C / Cl / CH_2CH_3 — R
    (4) 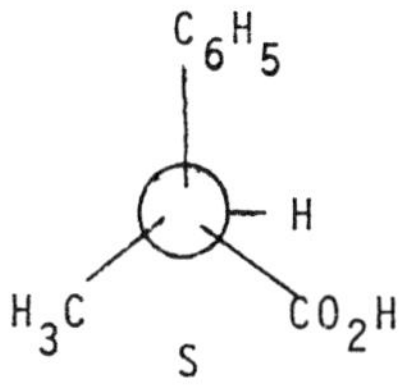 C_6H_5 / H / H_3C / CO_2H — S

20. (1) 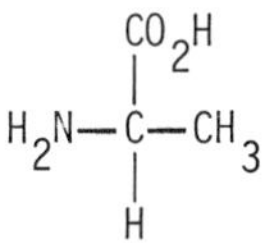 $H_2N-\underset{H}{\underset{|}{C}}(CO_2H)-CH_3$
    (2) $HOCH_2-\underset{H}{\underset{|}{C}}(CO_2H)-NH_2$ 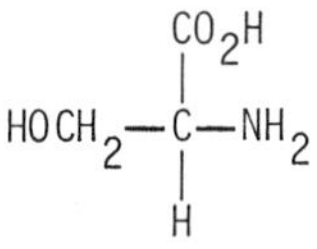
    (3) 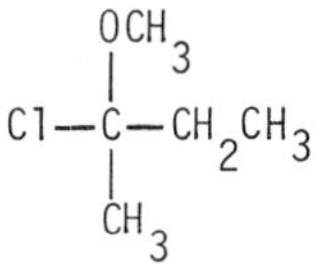 $Cl-\underset{CH_3}{\underset{|}{C}}(OCH_3)-CH_2CH_3$
    (4) 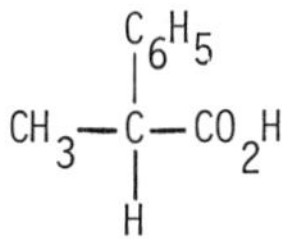 $CH_3-\underset{H}{\underset{|}{C}}(C_6H_5)-CO_2H$

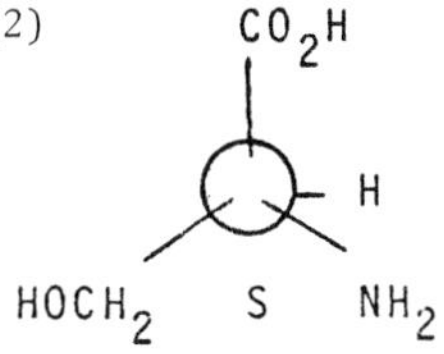

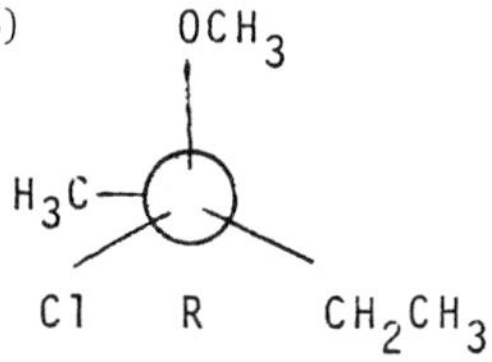

21. (1) E, $C_6H_5 > H$; $CO_2H > H$ (2) Z, $C_6H_5 > CH_3$, C—C versus C—H; Cl > F (3) Z, $CH_3CH_2 > CH_3$; $CH_3CH_2CH_2 > CH_3CH_2$
   (4) E, $HSCH_2 > HOCH_2$; $(CH_3)_2CH > CH_3CH_2$

22. [Newman and wedge projections]

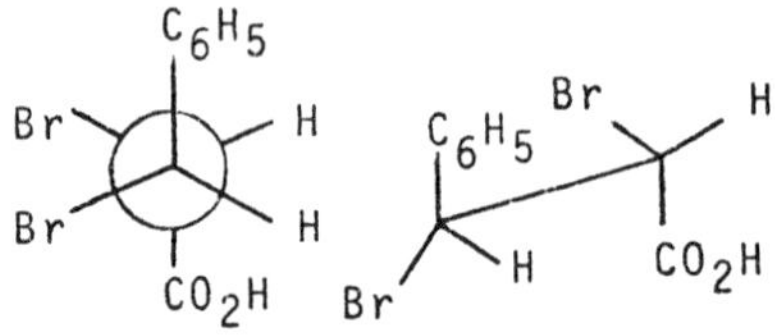

   threo, one enantiomer 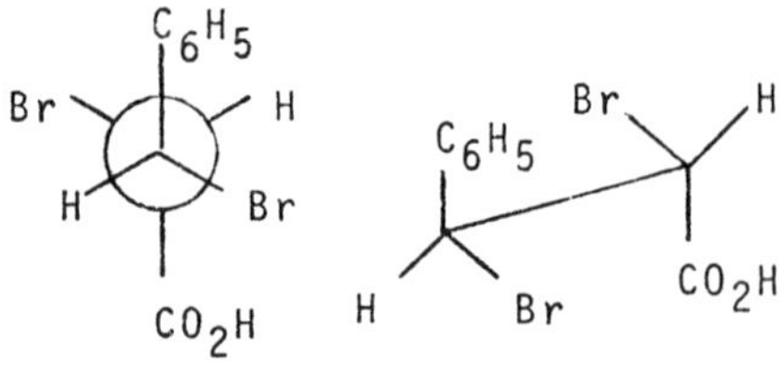   erythro, one enantiomer

23. (1) $\underline{2}$ $CH_3\!\!>\!\!C\!=\!C\!\!<\!\!CO_2H \atop H$, $\quad$ $CH_3\!\!>\!\!C\!=\!C\!\!<\!\!H \atop CO_2H$

   (2) $\underline{4}$

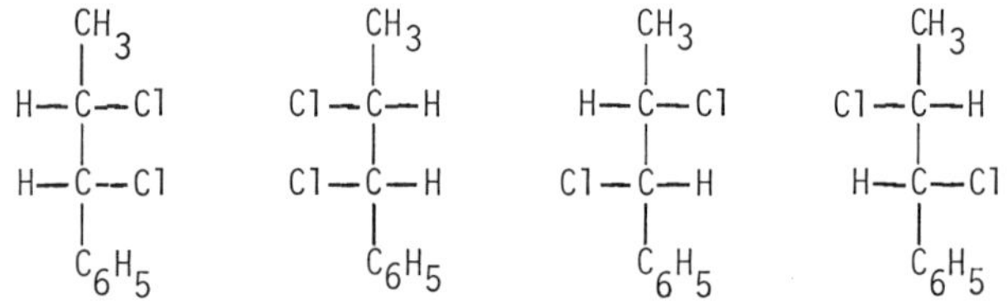

CH_3	CH_3	CH_3	CH_3
H—C—Cl	Cl—C—H	H—C—Cl	Cl—C—H
H—C—Cl	Cl—C—H	Cl—C—H	H—C—Cl
C_6H_5	C_6H_5	C_6H_5	C_6H_5

   (3) $\underline{8}$

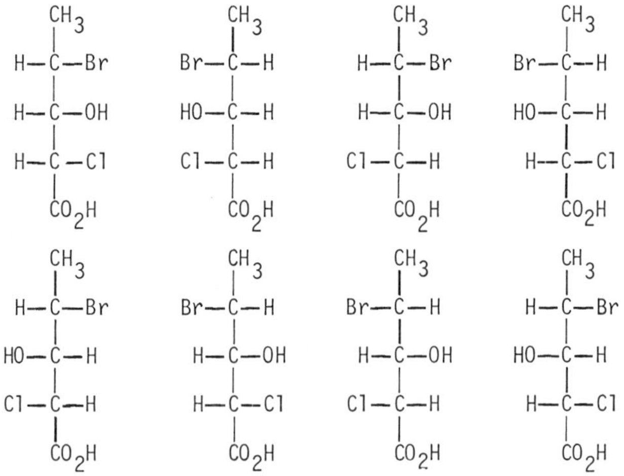

CH_3	CH_3	CH_3	CH_3
H—C—Br	Br—C—H	H—C—Br	Br—C—H
H—C—OH	HO—C—H	H—C—OH	HO—C—H
H—C—Cl	Cl—C—H	Cl—C—H	H—C—Cl
CO_2H	CO_2H	CO_2H	CO_2H

CH_3	CH_3	CH_3	CH_3
H—C—Br	Br—C—H	Br—C—H	H—C—Br
HO—C—H	H—C—OH	H—C—OH	HO—C—H
Cl—C—H	H—C—Cl	Cl—C—H	H—C—Cl
CO_2H	CO_2H	CO_2H	CO_2H

   (4) $\underline{2}$ $C_6H_5\!\!>\!\!C\!=\!C\!\!<\!\!H \atop C_6H_5 \atop H}$, $\quad$ $C_6H_5\!\!>\!\!C\!=\!C\!\!<\!\!C_6H_5 \atop H \atop H}$

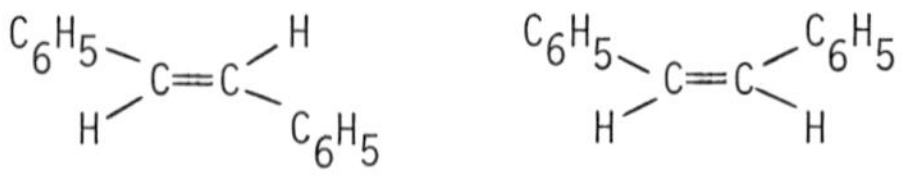

(5) 3

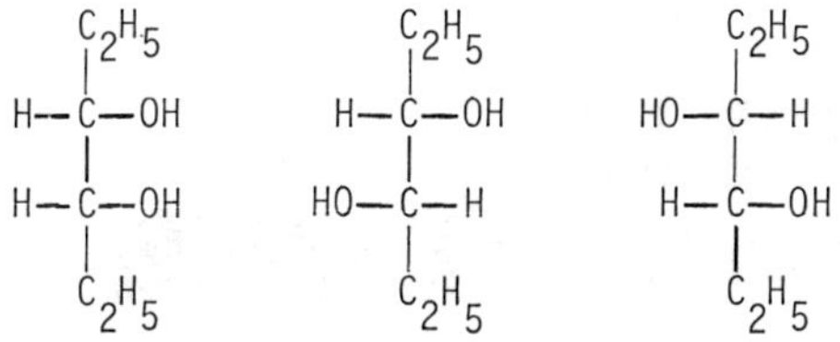

$\underline{3}$

H—C(C$_2$H$_5$)—OH
H—C(C$_2$H$_5$)—OH

meso

H—C(C$_2$H$_5$)—OH
HO—C(C$_2$H$_5$)—H

HO—C(C$_2$H$_5$)—H
H—C(C$_2$H$_5$)—OH

24. $CH_3CH_2C\equiv CCH_2CH_3 \xrightarrow{H_2, Pd-C}$ $\underset{H}{\overset{C_2H_5}{>}}C=C\underset{H}{\overset{C_2H_5}{<}}$

$\xrightarrow[Na, NH_3]{[H]}$ $\underset{H}{\overset{C_2H_5}{>}}C=C\underset{C_2H_5}{\overset{H}{<}}$

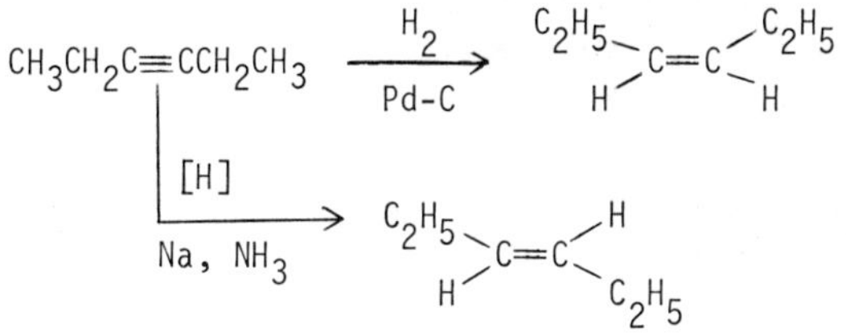

25.

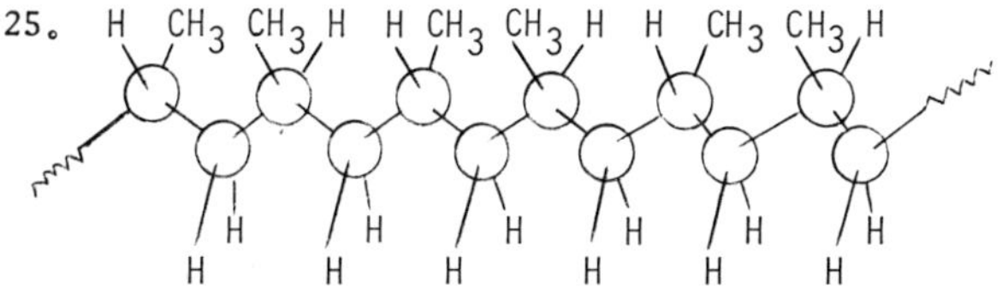

The H's and CH$_3$'s alternate; this is called syndiotactic.

26. (CH$_3$)$_2$NCH$_2$CH$_2$O—C$_6$H$_4$—C(=C(C$_6$H$_5$)(CH$_2$CH$_3$))—C$_6$H$_5$

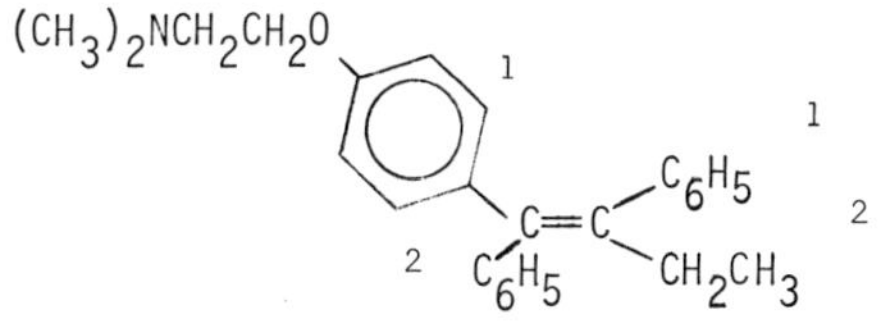

tamoxifen

Note that in the preferred name given, the parent chain has only two C's as compared with the alternative names with four C's. The reason it is preferred is that the priority of the principal function (amino over alkene) is overriding.

Priorities of the groups are designated 1,2 on the two C atoms. Thus, the configuration is $\underline{Z}$. Note that a <u>trans</u> designation does not necessarily correspond to E.

chapter 7

ALICYCLIC HYDROCARBONS AND DERIVATIVES: THE STEROIDS

1.

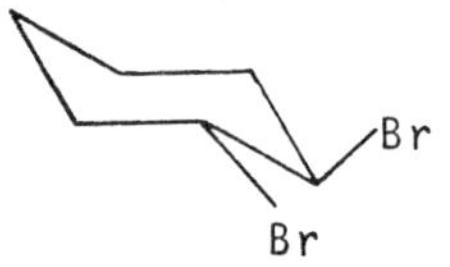

   trans-1,2-dibromo-
   cyclohexane

   cis-1,2-dibromo-
   cyclohexane

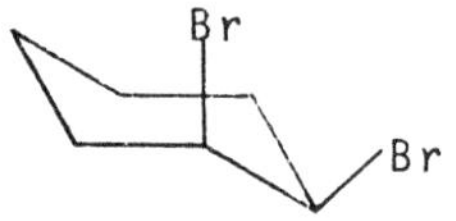

2.

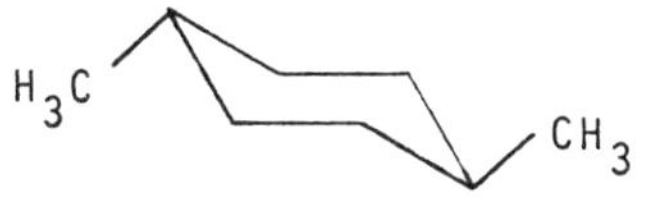

   Neither of these 1,4-dimethylcyclohexanes is an optical isomer because a plane of symmetry would pass through the 1,4- positions:

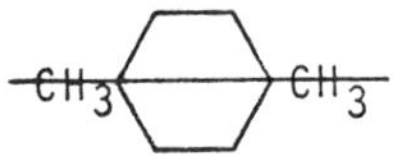

   ~~CH₃~~ ~~CH₃~~ (with hexagon)

3. 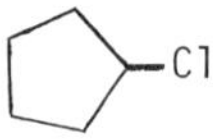 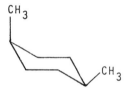—Cl, $CH_3CH_2CH_2CH_2CH_2Cl$, $CH_3CH_2CH_2\underset{Cl}{C}HCH_3$, $CH_3CH_2\underset{Cl}{C}HCH_2CH_3$

   The fact that only one monochlorination product would be obtained in the chlorination of cyclopentane versus three for n-pentane suggests that the former case is of greater preparative value.

4. The hydroxyl-bearing carbons of both isomers are similar asymmetric carbon atoms. Both have the same four points of attachment:

   $-OH$, $H-$, $-CH_2-$, $-\underset{OH}{C}H-$

   The trans isomer, therefore, would constitute a pair of enantiomers, while the cis isomer corresponds to a single optically inactive meso isomer (Section 6.9).

5. $CH_3CHCH_2CH_2Br \xrightarrow{Zn} CH_3-\triangle + ZnBr_2$
   $\quad\quad |$
   $\quad\;\; Br$

   $CH_3CH-CHCH_3 \xrightarrow{Zn} CH_3CH=CHCH_3 + ZnBr_2$
   $\quad\;\; |\quad\;\; |$
   $\quad\; Br\;\;\, Br$

6. (1) $CH_3CH_2CH_2OSO_2OH$ (2) $CH_3CH_2CH_2Br$

7. $\square \xrightarrow[120°]{H_2,\ Ni} CH_3CH_2CH_2CH_3$

8. $\square\!-\!Br \xrightarrow[Zn,\ HCl-CH_3CO_2H]{[H]} \square$

9.
$\begin{array}{c} HC\!\!\nearrow^{\!\!CH_2} \\ | \\ HC\!\!\searrow_{\!\!CH_2} \end{array} + \begin{array}{c} H_2C\!\!\nwarrow^{\!\!CH} \\ | \\ H_2C\!\!\swarrow_{\!\!CH} \end{array} \xrightarrow[TiCl_4]{(C_2H_5)_3Al} \bigcirc$

10.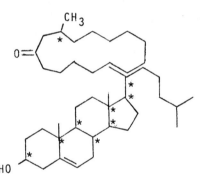

       1 asymmetric carbon atom
       2 optical isomers

       8 asymmetric carbon atoms
       256 theoretically possible optical isomers

11. naphthalene $\xrightarrow[heat]{H_2,\ Ni}$ trans-decalin   cis-decalin

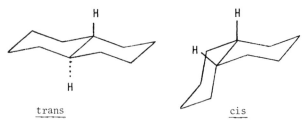

Optical isomers do not exist for either of these geometric isomers.

12. C_8H_8

 cyclooctatetraene

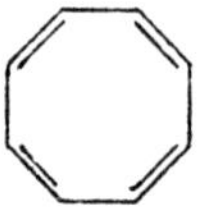

$C_{10}H_{22}$

$C_3(CH_2)_8CH_3$ decane

mp $-27°$, bp $142-143°$

mp $-30°$, bp $174°$

13. (1) C_6H_5

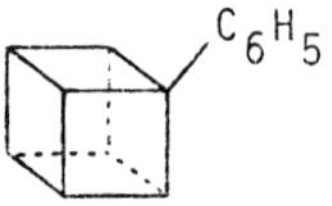

(2)

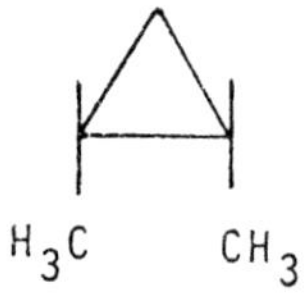

$H_3C \quad CH_3$

(3)

Cl

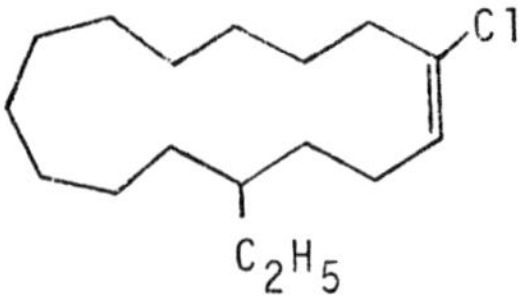

cis refers to the C=C in this case.

C_2H_5

(4)

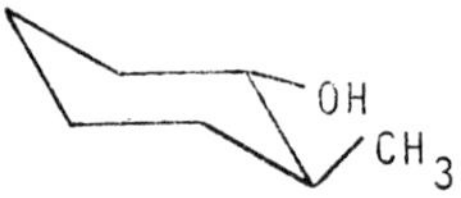

OH
CH_3

(5)

14. (1)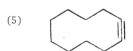

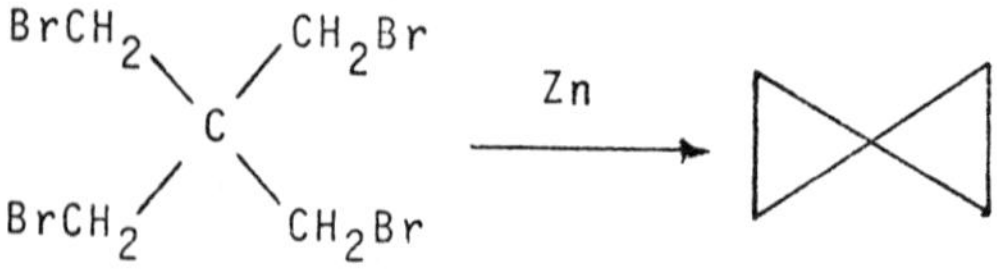

$BrCH_2$ CH_2Br
  C
$BrCH_2$ CH_2Br

Zn →

spiropentane

+ $ZnBr_2$

(2) 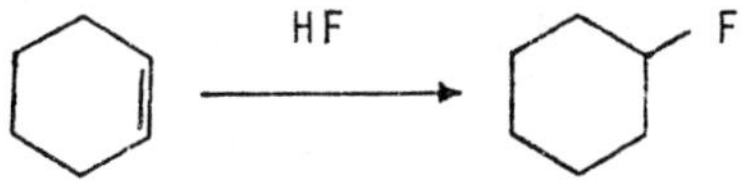 cyclohexene + HF → fluorocyclohexane Hydrogen fluoride cannot be added to all double bonds; it sometimes causes polymerization of an alkene.

(3)  CH_3–cyclopropane (with δ⁻) + HBr → $CH_3CHCH_2CH_3$ with Br on middle carbon The ring opens in such a way as to give a <u>secondary</u> carbonium ion.

(4) 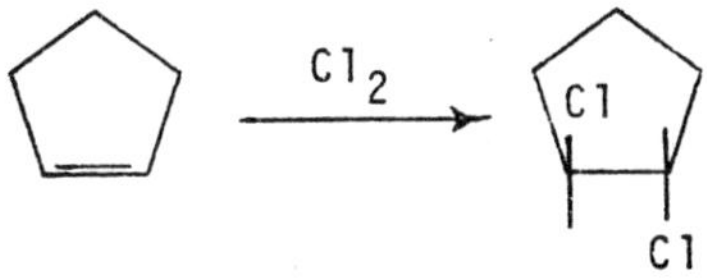 cyclopentene + Cl_2 → 1,2-dichlorocyclopentane (Cl, Cl)

(5) 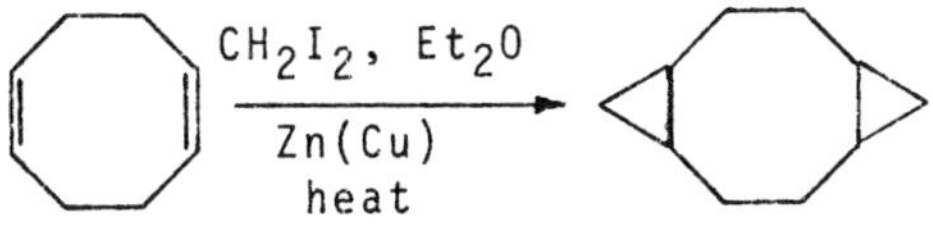 cyclooctadiene + CH_2I_2, Et_2O, Zn(Cu), heat → bis-cyclopropanated product

15. 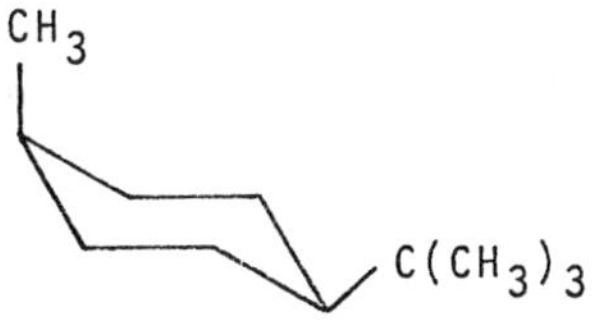 1-methyl-4-tert-butylcyclohexane (chair form) with $C(CH_3)_3$ equatorial The bulky <u>tert-butyl</u> group must occupy an equatorial position in preference to methyl group.

16.
(1) a. $CH_3CH=CH_2 + H_2O + [O]$ $\xrightarrow{KMnO_4}$ CH_3CHCH_2OH with OH on middle carbon

Lavender color disappears; cyclopropane does not react.

(2) b. 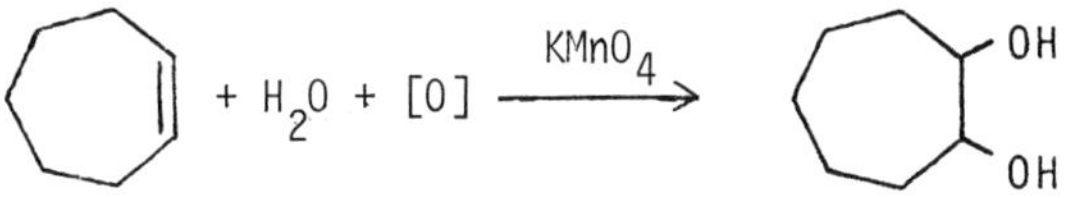 cycloheptene $+ H_2O + [O]$ $\xrightarrow{KMnO_4}$ cycloheptane-1,2-diol (OH, OH)
<u>cis</u> isomer

As in (1), cycloheptane does not react.

(3) b. 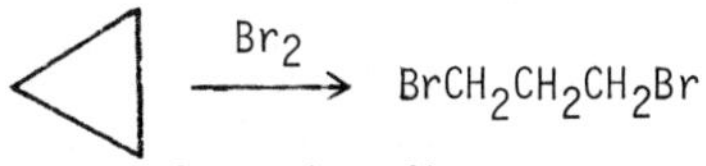 cyclopropane $\xrightarrow{Br_2}$ $BrCH_2CH_2CH_2Br$
Red color of bromine disappears; cyclobutane does not react.

17.

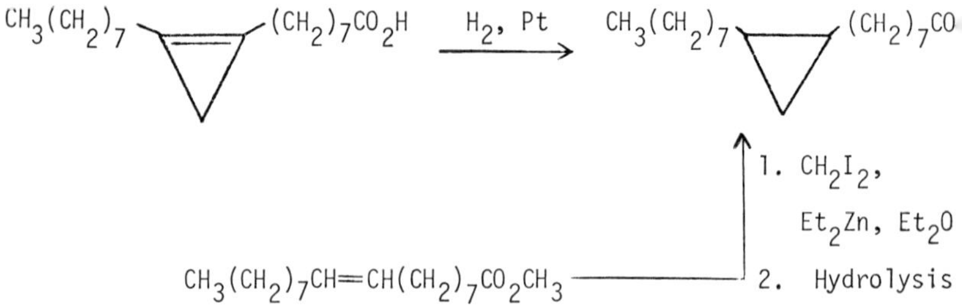

$CH_3(CH_2)_7$–[cyclopropane]–$(CH_2)_7CO_2H$ $\xrightarrow{H_2, Pt}$ $CH_3(CH_2)_7$–[cyclopropane]–$(CH_2)_7CO_2H$

↑ 1. CH_2I_2, Et_2Zn, Et_2O
2. Hydrolysis

$CH_3(CH_2)_7CH=CH(CH_2)_7CO_2CH_3$

Hydrogenation at room temperature with Pt catalyst avoids opening of the ring.

18. (1) 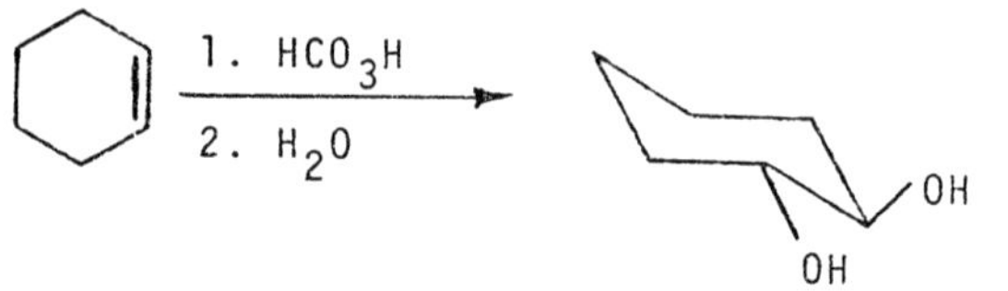 cyclohexene $\xrightarrow{\text{1. } HCO_3H,\ \text{2. } H_2O}$ trans-1,2-cyclohexanediol (OH, OH) Predominantly trans.

(2) muscone or civetone (See 7.15 for formulas.)

(3) 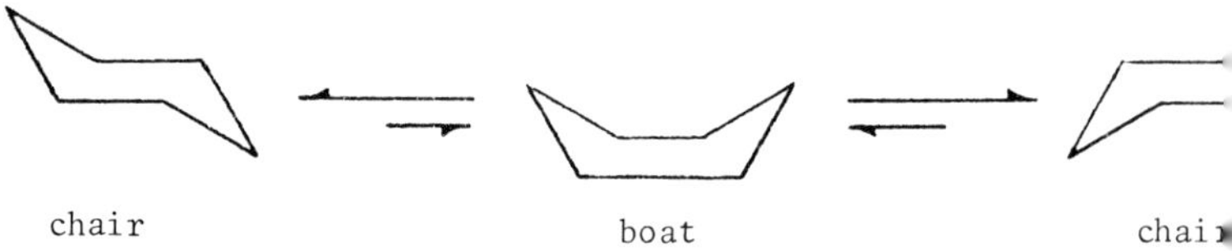 chair ⇌ boat ⇌ chair

(4) See Pb. 7.9.

(5)  [cyclohexane chair with]–$C(CH_3)_3$

19. (1) 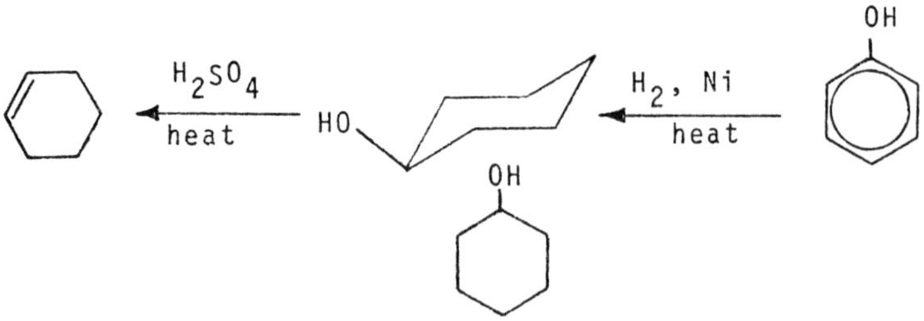 cyclohexene $\xleftarrow{H_2SO_4,\ \text{heat}}$ trans-1,2-cyclohexanediol (HO, OH) $\xleftarrow{H_2,\ Ni,\ \text{heat}}$ phenol (OH)

We should become accustomed to using both conformational and hexagonal formulas for cyclohexane; often the conformational formulas are favored. However, the conformational formula of cyclohexene having a flat side is less readily visualized, and, hence, we will retain the line formula in this case.

(2) 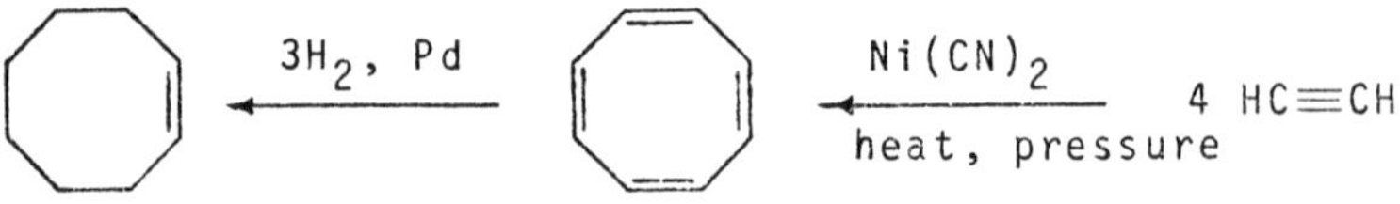 cyclooctane $\xleftarrow{3H_2,\ Pd}$ cyclooctatetraene $\xleftarrow{Ni(CN)_2,\ \text{heat, pressure}}$ 4 $HC\equiv CH$

(3) (3) 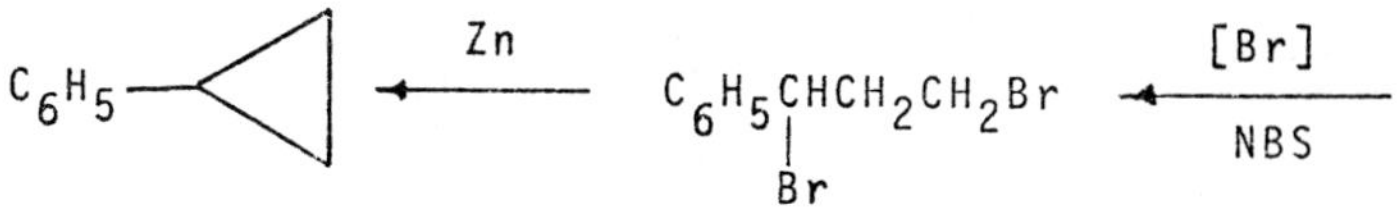 C_6H_5—△ $\xleftarrow{Zn}$ $C_6H_5\underset{Br}{CHCH_2CH_2Br}$ $\xleftarrow{[Br]}{NBS}$

$C_6H_5CH_2CH_2CH_2Br$

Although bromine can also be employed, NBS (N-bromosuccinimide) is more effective than bromine in replacing a hydrogen on carbon adjacent to a benzene ring; NBS has the structure:

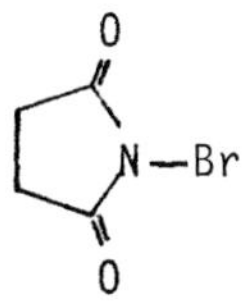

(structure of N-bromosuccinimide: five-membered ring with two C=O groups flanking N—Br)

chapter 8

ARENES: BENZENE-TYPE HYDROCARBONS

1. (1) 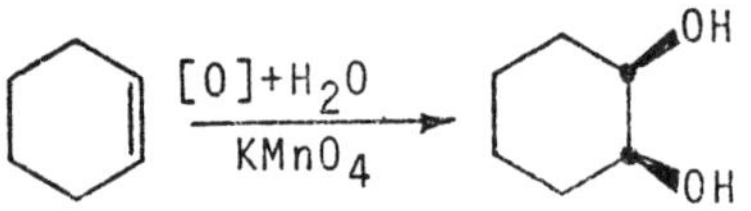 cyclohexene $\xrightarrow{[O]+H_2O \atop KMnO_4}$ cyclohexane-1,2-diol (cis)

   (2) 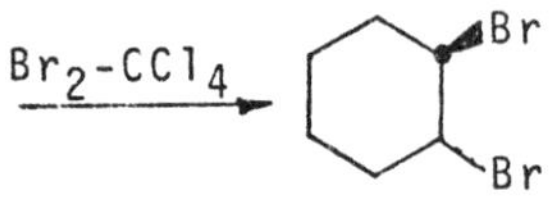 $\xrightarrow{Br_2-CCl_4}$ 1,2-dibromocyclohexane (trans)

   Compare with Pb. 7.2.

   (3) 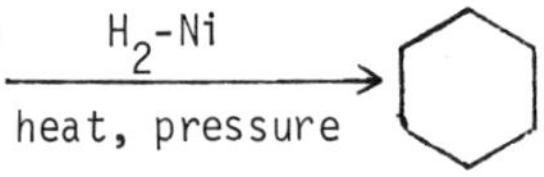 $\xrightarrow[\text{heat, pressure}]{H_2-Ni}$ cyclohexane

2. 1,2-diiodobenzene structures

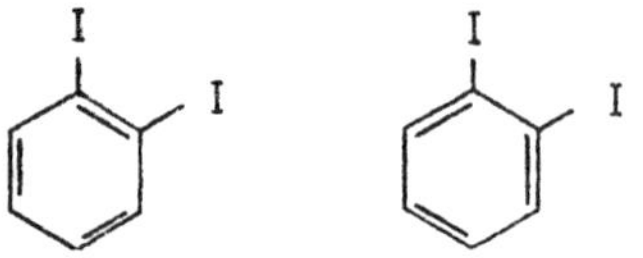

   Isomers could not be isolated because all bonds in benzene are identical as indicated by modern theory:

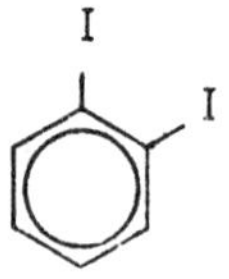

3. 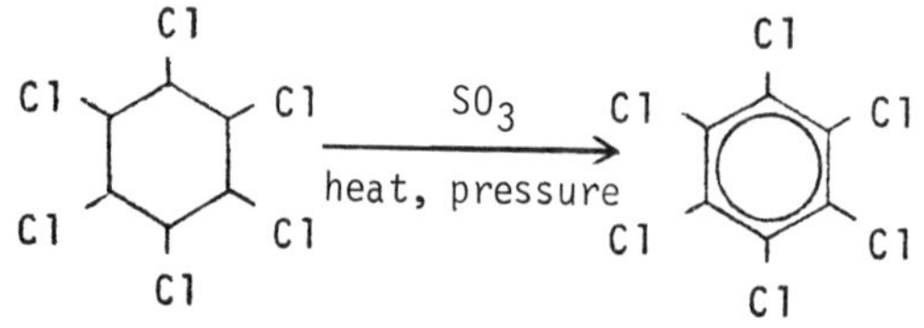 hexachlorocyclohexane $\xrightarrow[\text{heat, pressure}]{SO_3}$ hexachlorobenzene

4. 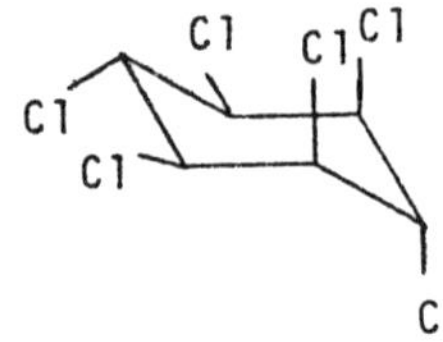 hexachlorocyclohexane (chair form)

5. 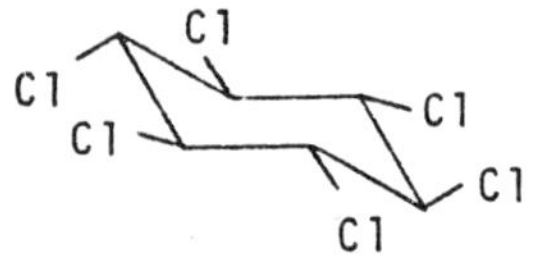 hexachlorocyclohexane (chair form)

Neither the γ nor β form is capable of existing as enantiomers, because they both would have a plane of symmetry through the center of the ring.

6.

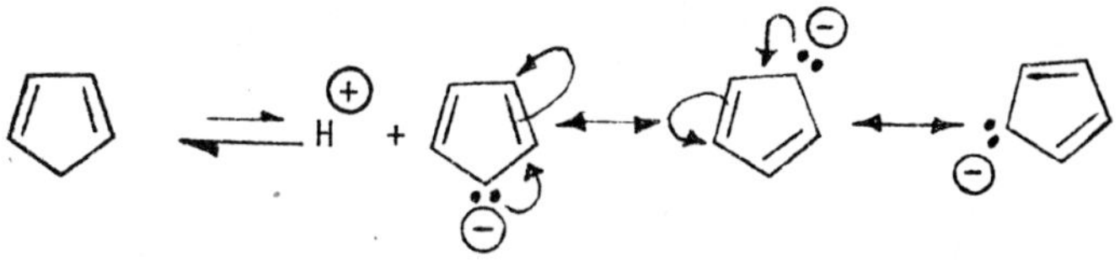

[resonance structures of cyclopentadienyl anion]

If a proton is removed from cyclopentadiene, the resulting anion with 6π electrons (4n + 2, where n = 1) has aromatic character and thus stability. Hence there is an increased tendency toward loss of a proton and enhanced acidity.

7. [cyclooctatetraene structures]   With 8π electrons, it does not meet Hückel rule requirements (see Table 8.1). Also, the 8-membered ring could pucker out-of-plane.

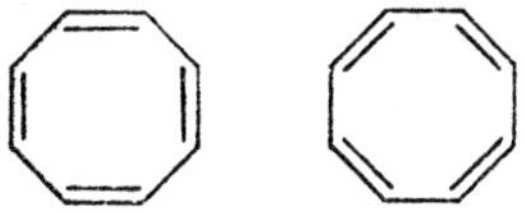

8.

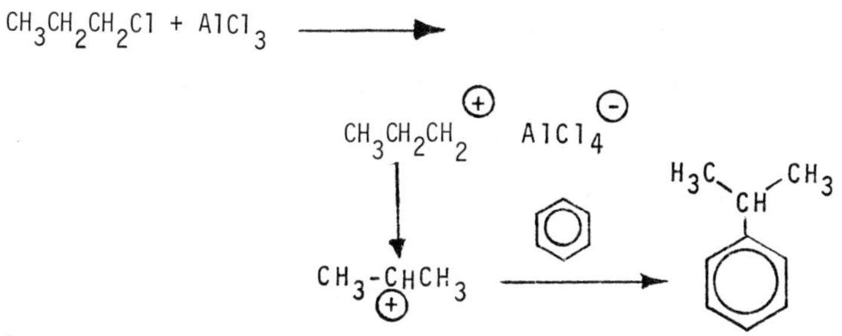

$CH_3CH_2CH_2Cl + AlCl_3 \longrightarrow$

$CH_3CH_2CH_2^{(+)}$ $AlCl_4^{(-)}$

$\downarrow$

$CH_3\text{-}\overset{(+)}{C}HCH_3$ + [benzene] $\longrightarrow$ [isopropylbenzene]

A <u>primary</u> carbonium ion rearranges to a more stable <u>secondary</u> carbonium ion.

9.

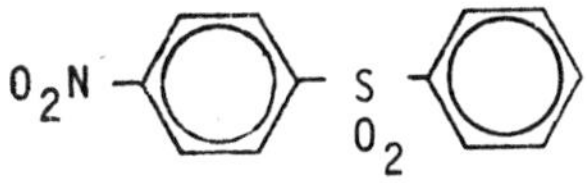

$O_2N\text{-}\bigcirc\text{-}\underset{O_2}{S}\text{-}\bigcirc$ This type of compound is known as a sulfone.

10.

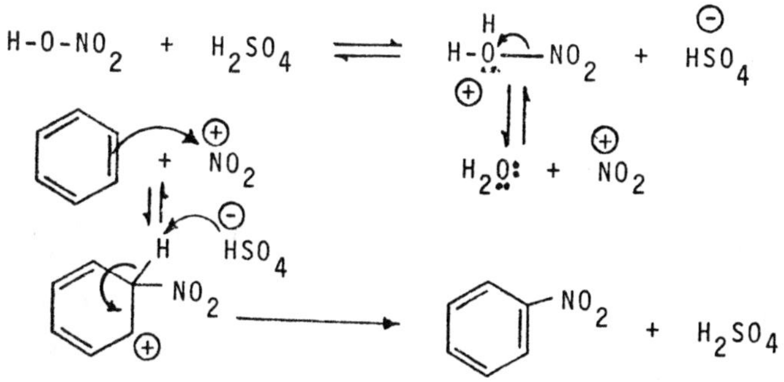

$H\text{-}O\text{-}NO_2 + H_2SO_4 \rightleftharpoons H\text{-}\overset{(+)}{\underset{H}{O}}\text{-}NO_2 + HSO_4^{(-)}$

$\updownarrow$

$H_2\overset{..}{O}\text{:} + \overset{(+)}{NO_2}$

[benzene + NO_2^+ → arenium intermediate with H, HSO_4^-, NO_2 → nitrobenzene + H_2SO_4]

11.

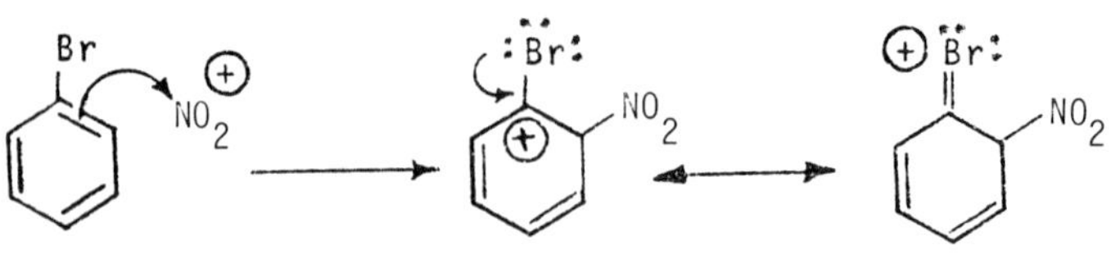

Carbonium ion (1) is particularly stable because it can be delocalized to (2) by interaction of a nonbonding pair of electrons from bromine.

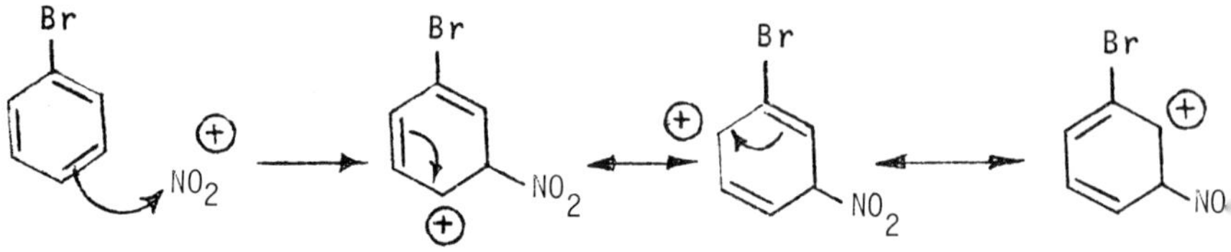

Because it is impossible to extend the positive charge to the bromine, meta substitution is far less favorable than ortho or para.

12. If a positive charge is present or can be developed on the atom to which the benzene ring is attached, the group will be meta-directing. This is so because only with meta substitution is it possible to avoid a highly unstable carbonium ion with positive charges on adjacent atoms.

(1) Positive charge clearly present on N.

(2) 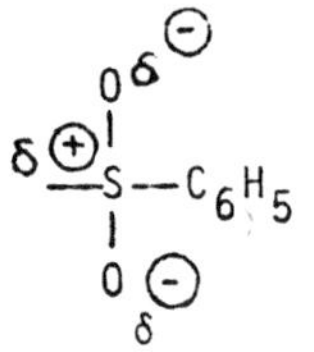 $\overset{\delta+}{S}-C_6H_5$ with O$^{\delta-}$ above and O$^{\delta-}$ below

(3) 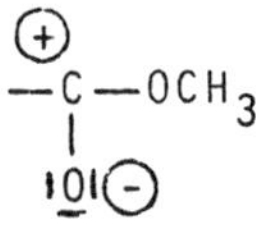 $-\overset{+}{C}-OCH_3$ with $:\underline{O}:^{\ominus}$

(4) 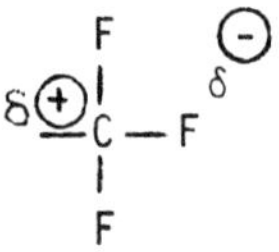 $\overset{\delta+}{C}-F^{\delta-}$ with F above and F below

13. (1) 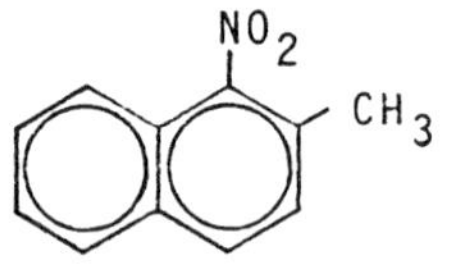 1-NO$_2$, 2-CH$_3$ naphthalene   (2) 2-methyl-3,5-dinitrobenzonitrile

(3) 1-chloro-8-iodonaphthalene   (4) 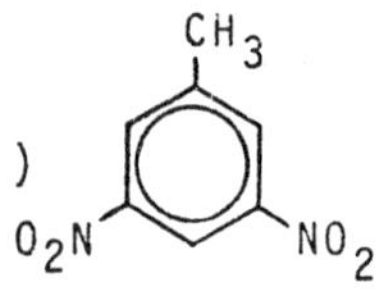 1-CH$_3$, 3-O$_2$N, 5-NO$_2$ benzene

(5) 9-methylphenanthrene   (6) 2,4-dichlorophenol

(7) 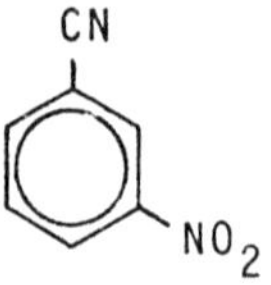 1-CN, 3-NO$_2$ benzene   (8) 2-chloro-6-nitrotoluene

(9) 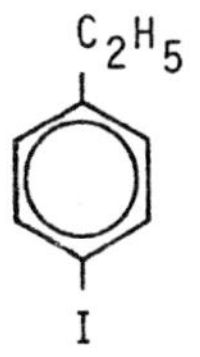 [ethyl-benzene with I para] C$_2$H$_5$ / I

(10) 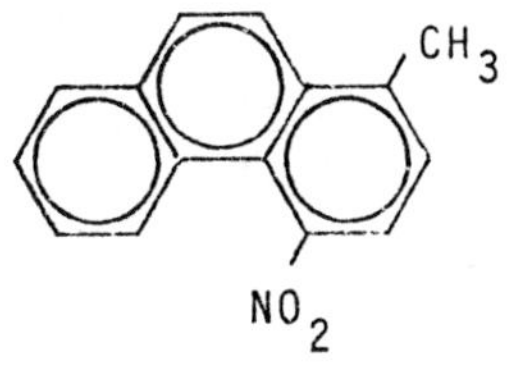 [phenanthrene with CH$_3$ and NO$_2$ substituents]

14.

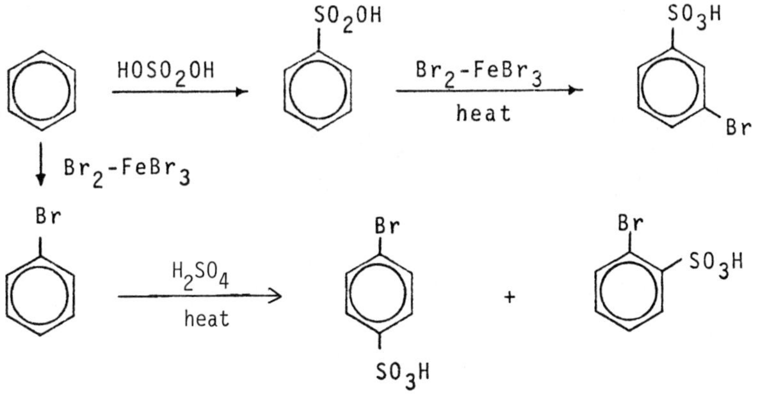

C_6H_6 →(HOSO$_2$OH)→ C$_6$H$_5$-SO$_2$OH →(Br$_2$-FeBr$_3$, heat)→ m-Br-C$_6$H$_4$-SO$_3$H

↓ Br$_2$-FeBr$_3$

C$_6$H$_5$Br →(H$_2$SO$_4$, heat)→ p-Br-C$_6$H$_4$-SO$_3$H + o-Br-C$_6$H$_4$-SO$_3$H

15.

$H\text{-}\ddot{O}\text{-}\underset{O_2}{S}\text{-}OH + H_2SO_4 \rightleftharpoons H\text{-}\overset{+}{O}(H)\text{-}\underset{O_2}{S}\text{-}OH + HSO_4^-\rightleftharpoons$

$H_2O + SO_2OH^+ \rightleftharpoons SO_3^- + H_3O^+$

[toluene + SO$_3$ → electrophilic addition → arenium ion intermediate + HSO$_4^-$ → ortho-methylbenzenesulfonate + H$_2$SO$_4$]

[ortho-CH$_3$-C$_6$H$_4$-SO$_3^-$ + H$_3$O$^+$ ⇌ ortho-CH$_3$-C$_6$H$_4$-SO$_3$H + H$_2$O]

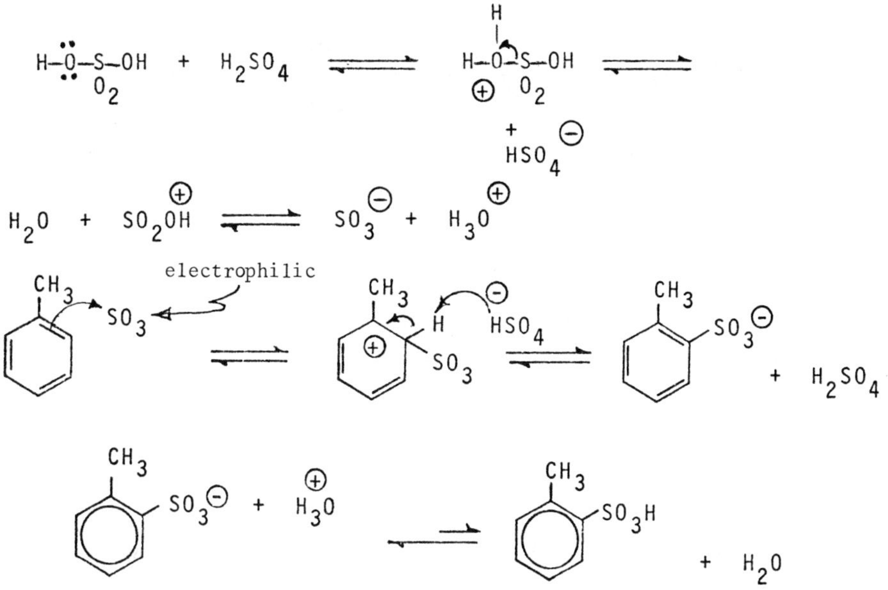

Sulfur trioxide, a Lewis acid because of a sextet of electrons about sulfur, is the electrophilic species. Either <u>ortho</u> or <u>para</u> substitution occurs, as the positive charge at the <u>site of the</u> methyl substituent is stabilized by the methyl group's electron-donating character.

16.

(1) 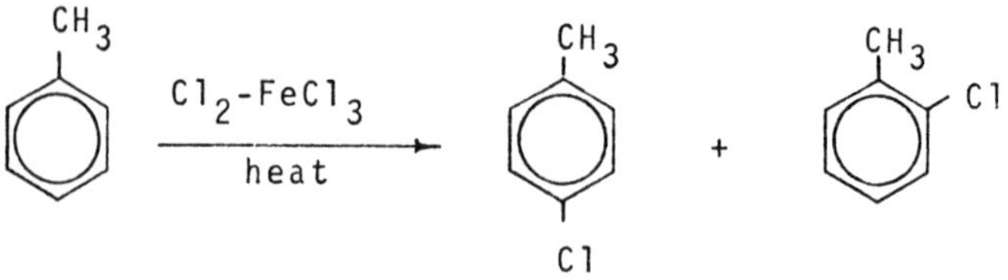
$$\text{CH}_3\text{-C}_6\text{H}_5 \xrightarrow[\text{heat}]{Cl_2-FeCl_3} \text{p-ClC}_6\text{H}_4\text{CH}_3 + \text{o-ClC}_6\text{H}_4\text{CH}_3$$

o,p-Isomers may be readily separated.

(2) 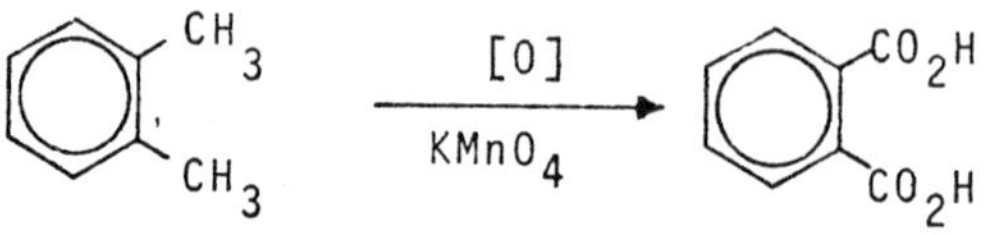
$$\text{o-C}_6\text{H}_4(\text{CH}_3)_2 \xrightarrow[\text{KMnO}_4]{[O]} \text{o-C}_6\text{H}_4(\text{CO}_2\text{H})_2$$

(3)
$$\text{C}_6\text{H}_5\text{CH}_3 \xrightarrow{HNO_3-H_2SO_4} 2,4,6\text{-(O}_2\text{N)}_3\text{C}_6\text{H}_2\text{CH}_3 \xrightarrow[\text{KMnO}_4]{[O]} 2,4,6\text{-(O}_2\text{N)}_3\text{C}_6\text{H}_2\text{CO}_2\text{H}$$

(4) benzene + cyclohexene $\xrightarrow{AlCl_3}$ phenylcyclohexane

(5)
$$\text{o-}(CH_3)_2C_6H_4 \xrightarrow[\text{heat}]{H_2SO_4} CH_3\text{-C}_6H_3(CH_3)\text{-SO}_3H \text{ (para)} + CH_3\text{-C}_6H_3(CH_3)\text{-SO}_3H \text{ (ortho)}$$

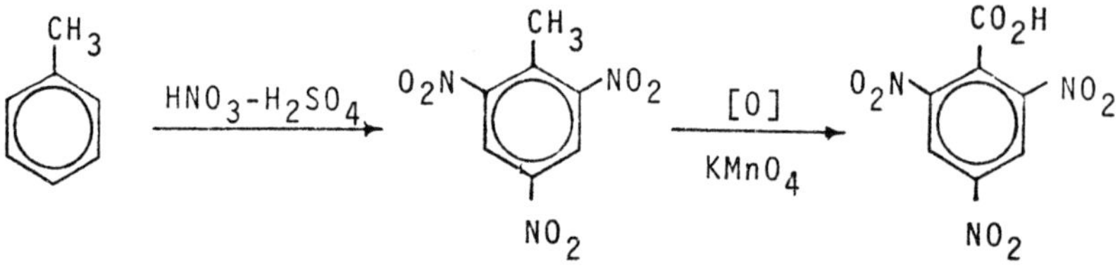

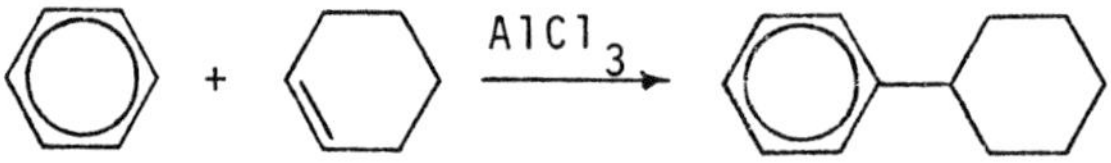

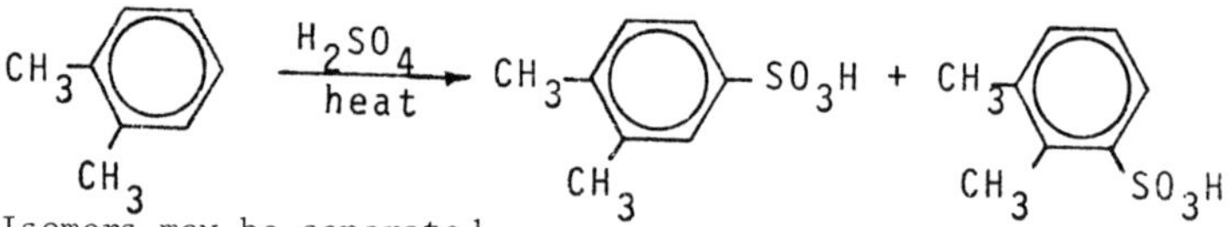

Isomers may be separated.

(6) 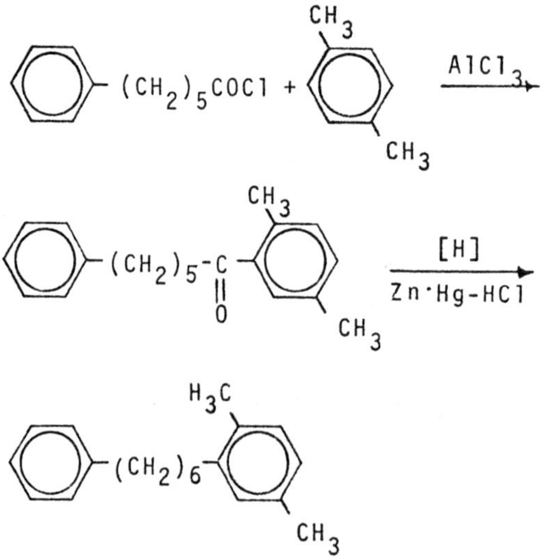
$$C_6H_5\text{-(CH}_2)_5COCl + 2,4\text{-(CH}_3)_2C_6H_4 \xrightarrow{AlCl_3}$$

$$C_6H_5\text{-(CH}_2)_5\text{-CO-}C_6H_3(CH_3)_2 \xrightarrow[\text{Zn·Hg-HCl}]{[H]}$$

$$C_6H_5\text{-(CH}_2)_6\text{-}C_6H_3(CH_3)_2$$

17.

(1) 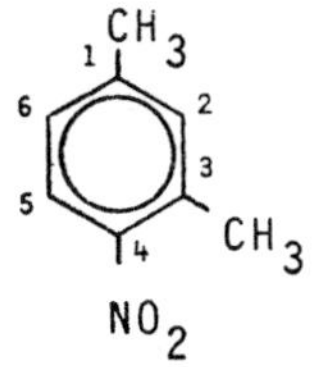 [structure: benzene with CH₃ at position 1, CH₃ at position 4, NO₂ at position 4; positions numbered 1-6]

Nitration would occur in equivalent positions 4 and 6 ortho and para to the methyl groups. Nitration in position 5 would not occur because it is meta to the methyl groups. Nitration in position 2 between substituents does not occur because of steric blocking, unless it is the only possibility (Pb. 8.17.4).

(2) 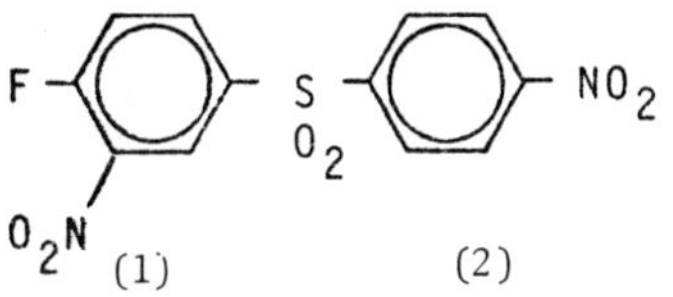 [structures: ring (1) with F and O_2N substituents; ring (2) with SO_2 linker and NO_2]

Nitration does not occur in ring (2) because it is too highly deactivated relative to ring (1) by the two meta-directing groups. Because F is ortho-, para-directing and $-SO_2-$ meta-directing, the logical position in ring (1) is assumed by the nitro group.

(3) 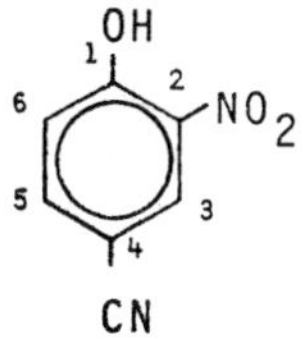 [structure: benzene with OH at position 1, NO₂ at position 2, CN at position 4]

Equivalent positions 2 and 6 are reinforced as logical locations in respect to ortho-directing OH and meta-directing CN.

(4) 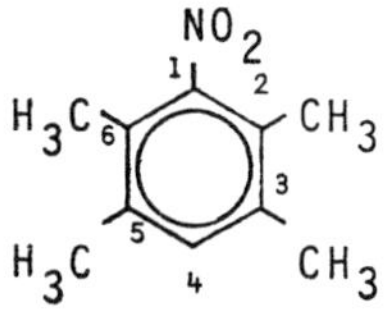 [structure: benzene with NO₂ at position 1 (2?), CH₃ groups at positions 2, 3, 5, 6; H₃C labels]

Highly activated by the four methyl groups, mono nitration occurs with great ease in equivalent positions 1 and 4 despite steric hindrance.

(5) 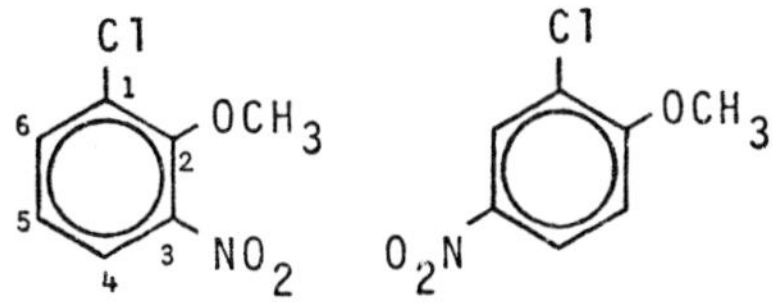 [two structures: benzene with Cl at 1, OCH₃ at 2, NO₂ at 3; and benzene with Cl, OCH₃, O₂N]

87

(6) Nitration occurs primarily in positions 3 and 5 to give two isomers. The more powerful of the two o,p-directing groups has predominant influence.

[Structures: 5-NO$_2$-substituted acetanilide with COCH$_3$ at position 3 (numbered 1,2,3,4,5,6 with CH$_3$CONH at position 1); and the para-NO$_2$ isomer with O$_2$N– and –COCH$_3$ on the ring and CH$_3$CONH substituent]

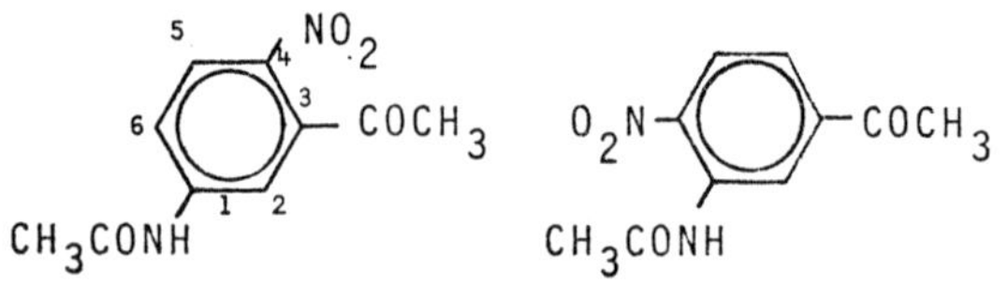

Two isomers are obtained with nitration occurring ortho and para to the acetamido group. An o,p-directing group always has predominant influence over an m-directing group.

18. Naphthalene (Table 3.5) has 10π electrons and phenanthrene 14, so that on the basis of the Hückel rule (Table 8.1) both would be expected to be aromatic. Azulene with 10π electrons would also be expected to be aromatic.

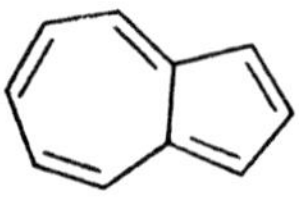

azulene

chapter 9

ORGANIC HALIDES AND NUCLEOPHILIC SUBSTITUTION: ORGANOMETALLICS

1. (1) $(CH_3)_3CCH_2Cl$ (2) 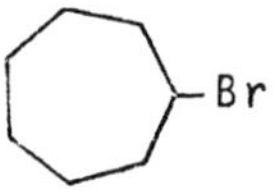 cycloheptyl–Br

2. (1) 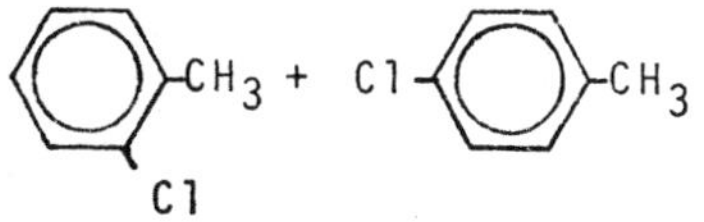 o-chlorotoluene + p-chlorotoluene   (2) 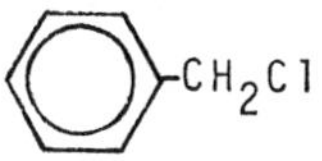 benzyl chloride (Ph–CH_2Cl)

3. (1) $CH_3CHClCH_2Cl$ (2) $CH_3CHClCH_3$

4. (1) 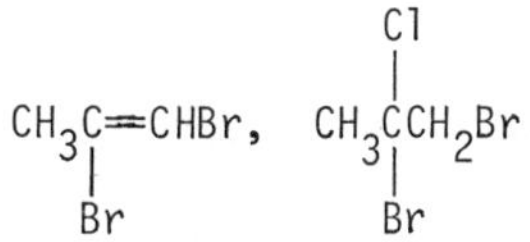 $CH_3C(Br)=CHBr$, $CH_3C(Cl)(Br)CH_2Br$

5. Ph–CH_2OH $\xrightarrow{HI}$ Ph–CH_2I

6. hexachlorobenzene $\xrightarrow[500°]{KF}$ hexafluorobenzene

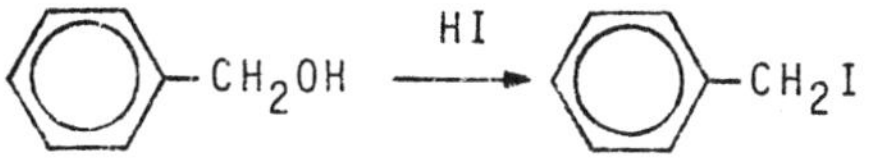

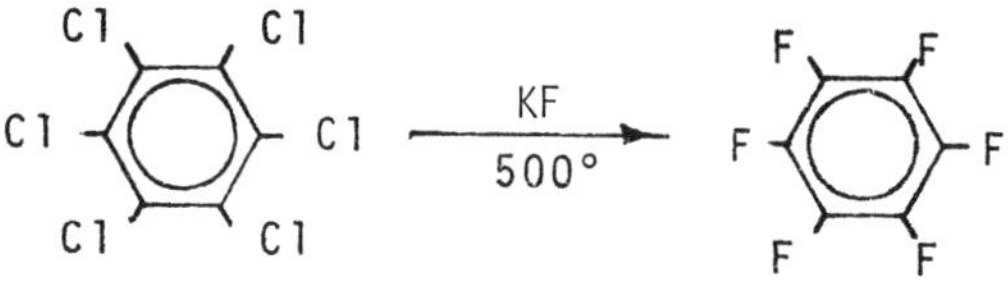

7. cyclohexene $\xrightarrow[HCCl_3,\ -75°]{Pb(OAc)_4-HF}$ trans-1,2-difluorocyclohexane

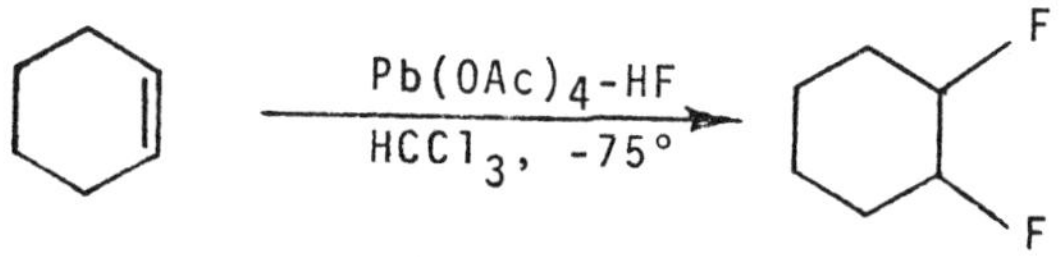

8. (1) $H_2C=CH_2 \xrightarrow{HF} CH_3CH_2F$ (2) CH_3CHFCH_3

9. $CH_3CH_2Br \xrightarrow{HgF_2} CH_3CH_2F$

10.
   (1) $CH_3CH_2CH_2CH=CHCH_2CH_3$ (2) 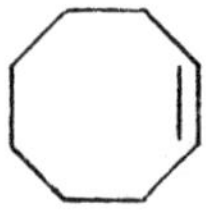 [cyclooctene structure]   (3) $C_6H_5C{\equiv}CH$

11. $C_6H_5-Cl \xrightarrow[(CH_3)_2CHOH-Mg]{[H]} C_6H_5-H + HCl$

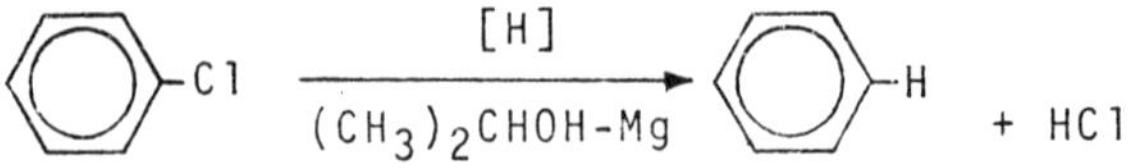

12. $(C_6H_5)_2CH_2 \xrightarrow{NaNH_2-NH_3 \text{ (liquid)}} (C_6H_5)_2CHNa + NH_3\uparrow$

13. $H_2C{=}CHCH_2Cl \rightleftharpoons [H_2C{\overset{\frown}{=}}CH{-}\overset{\oplus}{CH_2} \longleftrightarrow \overset{\oplus}{H_2C}{-}CH{=}CH_2]\overset{\ominus}{Cl}$

Delocalizatioh (and thus stability) of the cation is in effect as indicated by the two identical resonance formulas.

14. Both would exhibit first-order kinetics, because both can form carbonium ions which are resonance stabilized and would therefore follow the S_N1 mechanism.

15.
$O_2N{-}C_6H_3(NO_2){-}Cl \xrightarrow[\substack{125° \\ 2.\ \overset{+}{H}-H_2O}]{1.\ NaHCO_3-[\overset{\ominus}{OH}]} O_2N{-}C_6H_3(NO_2){-}OH$

$O_2N{-}C_6H_3(NO_2)(NO_2){-}Cl \xrightarrow[\text{heat}]{H_2O} O_2N{-}C_6H_3(NO_2)(NO_2){-}OH$

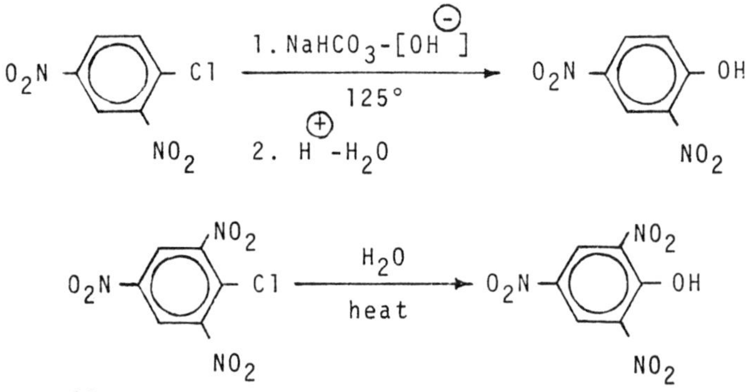

16. [Resonance structures of chloronitrobenzene showing electron delocalization through the nitro group]

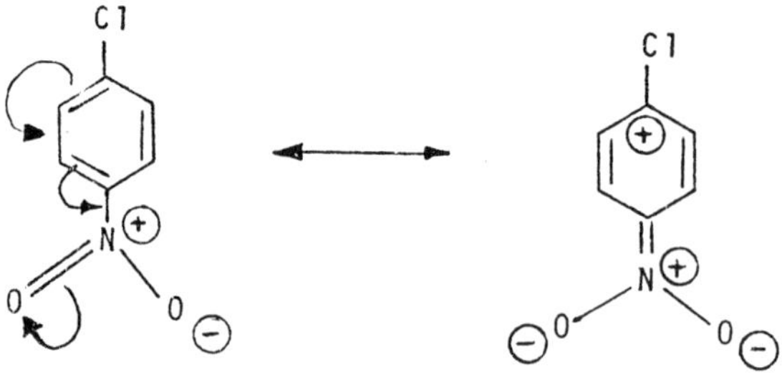

The C attached to the halogen is particularly electrophilic because the nitro group is electron-attracting.

17. [tetrahydrofuran structure]

18.

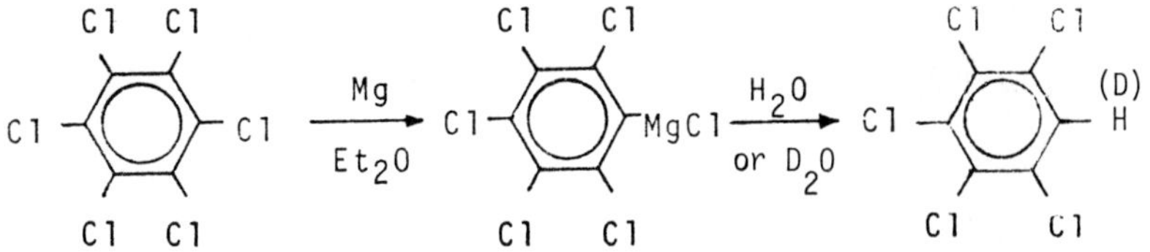

$$\text{C}_6\text{Cl}_5\text{-Cl} \xrightarrow{\text{Mg}, \text{Et}_2\text{O}} \text{C}_6\text{Cl}_5\text{-MgCl} \xrightarrow{\text{H}_2\text{O or D}_2\text{O}} \text{C}_6\text{Cl}_5\text{-H (D)}$$

(pentachlorobenzene ring with Cl at five positions, reacting with Mg in Et₂O to give the Grignard C₆Cl₅MgCl, then H₂O or D₂O to give C₆Cl₅H (D))

19.

$$2\text{CH}_3\text{Cl} \xrightarrow[\text{heat}]{\text{Si·Cu}} (\text{CH}_3)_2\text{SiCl}_2, \quad 4\text{CH}_3\text{Cl} \xrightarrow{\text{Si·Cu}} \text{CH}_3\text{SiCl}_3 + (\text{CH}_3)_3\text{SiCl}$$

$$4\text{C}_2\text{H}_5\text{Cl} \xrightarrow{\text{Pb·Na}} (\text{C}_2\text{H}_5)_4\text{Pb}$$

20.
(1) 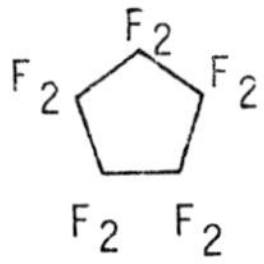 perfluorocyclopentane (cyclopentane with F₂ on each of the 5 carbons)

(2) 1-bromooctane, octyl bromide

(3) p-bromobenzyl bromide or p-α-dibromotoluene (CA)

(4) 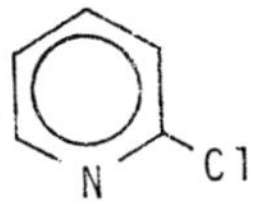 2-chloropyridine

(5) 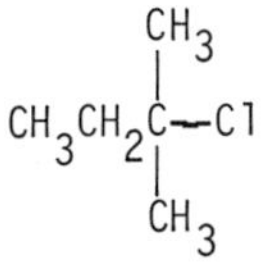 $\text{CH}_3\text{CH}_2\text{C(CH}_3)_2\text{Cl}$ (2-chloro-2-methylbutane structure shown as CH₃CH₂–C(CH₃)(CH₃)–Cl)

(6) 2,3,5-triiodobenzoic acid

(7) 1-bromo-2-fluorohexane

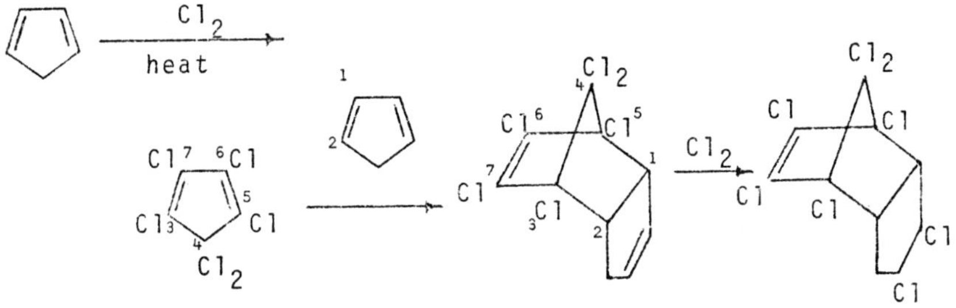

(8) 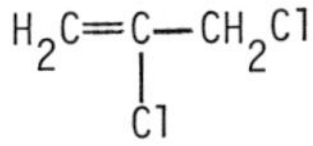 $\text{H}_2\text{C}=\text{C(Cl)}-\text{CH}_2\text{Cl}$

21. Cyclopentadiene $\xrightarrow{\text{Cl}_2, \text{heat}}$ hexachlorocyclopentadiene (C₅Cl₆, numbered 1–7) ; then Diels-Alder with cyclopentadiene gives intermediate; then Cl₂ adds to give chlordane.

22.
(1) 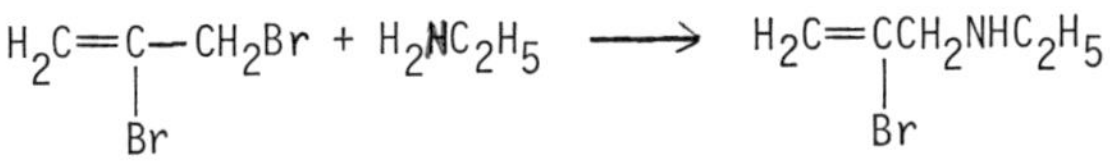 $\text{H}_2\text{C}=\text{C(Br)}-\text{CH}_2\text{Br} + \text{H}_2\text{NC}_2\text{H}_5 \longrightarrow \text{H}_2\text{C}=\text{C(Br)}-\text{CH}_2\text{NHC}_2\text{H}_5$

Br attached to doubly bonded C atom unreactive; Br attached to C atom once removed from C=C highly reactive. Same considerations hold for 4-chlorobenzyl chloride.

(2) 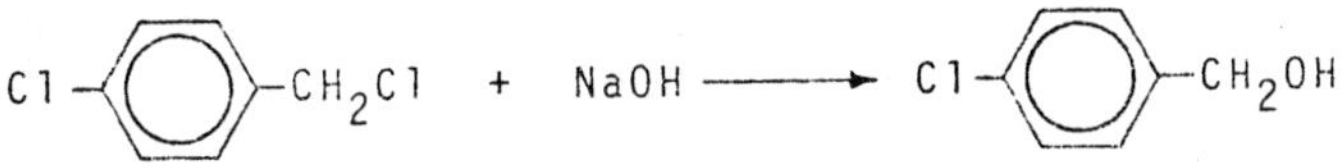 $\text{Cl-C}_6\text{H}_4\text{-CH}_2\text{Cl} + \text{NaOH} \longrightarrow \text{Cl-C}_6\text{H}_4\text{-CH}_2\text{OH}$

23.

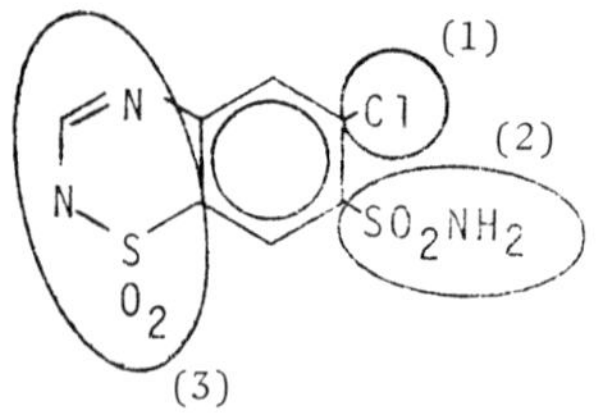

[Structure: benzene ring with Cl (1) and SO₂NH₂ (2) substituents, fused to a heterocyclic ring containing N=, N, S, O₂ (3)]

(1) A chloro substituent and a functional group of central interest in this chapter.   (2) Sulfonamide group—a group, which is similar to an amide group (Section 11.3.2).   (3) A heterocyclic group; 1,2,4-thiadiazine dioxide, two nitrogens, and a sulfur in the ring.

24.
(1) 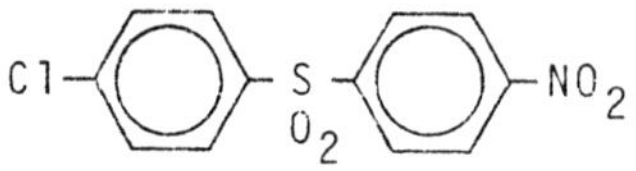 Cl—⟨C₆H₄⟩—S(O₂)—⟨C₆H₄⟩—NO₂

a sulfone

The reaction is carried out in dimethyl sulfoxide (CH_3SCH_3, DMSO, the simplest member of the sulfoxide series). $\overset{O}{\underset{\parallel}{}}$ DMSO is an aprotic (not a source of H⊕) solvent, frequently employed for nucleophilic substitution reactions. Much higher yields are obtained than when protic solvents such as ethanol are used.

(2) $(CH_3)_2CHCH_2CH_2CN$ (3) $C_6H_5\underset{\underset{C_6H_5}{|}}{CH}(CH_2)_5CH_3$

(4) 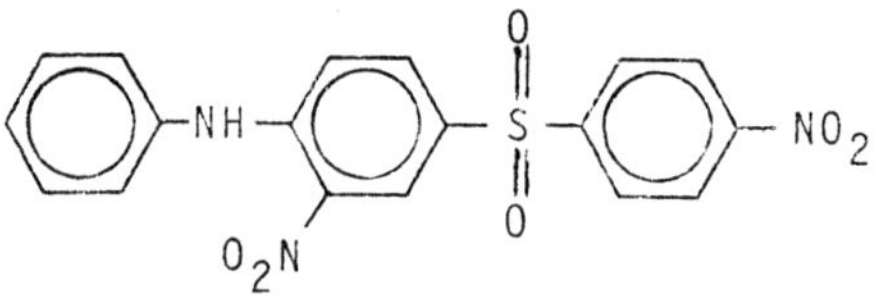 ⟨C₆H₅⟩—NH—⟨C₆H₃(NO₂)⟩—S(=O)(=O)—⟨C₆H₄⟩—NO₂

Another example of a sulfone. It is formed from the highly activated halodiphenyl sulfone in an aromatic nucleophilic substitution.

(5) $CH_3CH_2CH_2\underset{\underset{CH_3}{|}}{CH}-S-CH_2C_6H_5$

An organic sulfide is the product of this reaction.

(6) $C_6H_5CH_2CH_2\underset{\underset{C_6H_5}{|}}{CH}CO_2C_2H_5$

The methylene group of the starting material, ethyl phenylacetate, readily forms an anion which reacts with 1-bromo-2-phenylethane.

25. (1) 9.9   (2) 9.7   (3) 9.10   (4) 9.13   (5) 9.20   (6) 9.15

26.
$$HC\equiv CH \xrightarrow[BF_3]{2HF} CH_3CHF_2 \xrightarrow{heat} H_2C=CHF \xrightarrow[heat]{R\cdot} {-\!\!\left[CH_2CH\underset{F}{|}\right]\!\!-}_n \text{ Tedlar}$$

27. Methane is the main constituent of natural gas, which serves as the ultimate starting material for the synthesis of many important commercial chemicals. This problem emphasizes the importance of natural gas and petroleum as raw materials for the manufacture of many industrial chemicals and consumer products.

(1) $CH_4 \xrightarrow[h\nu]{3Cl_2} CHCl_3 \xrightarrow[SbF_5]{HF} CHClF_2 \xrightarrow{250°}$

$F_2C=CF_2 \xrightarrow[heat]{K_2S_2O_8} {-\!\!\left[CF_2CF_2\right]\!\!-}_n \quad n = \sim 10,000-35,000$

(2) $CH_4 \xrightarrow[h\nu]{4Cl_2} CCl_4 \xrightarrow[100°, pressure]{HF, SbF_5} CCl_2F_2$

(3) See (1) to $F_2C=CF_2 \xrightarrow[pressure]{200°} F_2\diamondsuit F_2$ (with F_2 on each corner)

(4) $CH_4 \xrightarrow[heat]{Cl_2} CH_3Cl \xrightarrow[heat]{Si\cdot Cu} (CH_3)_2SiCl_2 \xrightarrow{H_2O} {-\!\!\left[\underset{CH_3}{\overset{CH_3}{Si}}-O\right]\!\!-}_n$

28. (1)
$$K:\overset{+}{}\overset{-}{OC}(CH_3)_3 + HCCl_3 \longrightarrow :CCl_3^{-} + (CH_3)_3COH$$

$$\overset{(3)}{\longleftarrow} \overset{(2)\downarrow}{:CCl_2} + Cl^{-}$$

$$CH_3CH_2CH=CH_2$$

CH_3CH_2—(cyclopropane with Cl, Cl)

29. (1) $(CH_3)_3CMgBr + C_2H_5OH \longrightarrow (CH_3)_3CH + C_2H_5OMgBr$.

Like HOH, C_2H_5OH has a reactive H which will substitute for MgBr in a Grignard reagent.

(2) $CH_3CH_2\underset{CH_3}{\overset{CH_3}{\underset{|}{\overset{|}{C}}}}-MgBr + HC\equiv CH \longrightarrow CH_3CH_2\underset{CH_3}{\overset{CH_3}{\underset{|}{\overset{|}{C}}}}-H + HC\equiv CMgBr$

Acetylene also has reactive H's.

chapter 10

ALCOHOLS, PHENOLS, ETHERS, AND RELATED SULFUR COMPOUNDS

1. The S—H bond in ethanethiol is not as polar as the O—H bond in ethanol. Accordingly, ethanethiol will have a lower water solubility because it does not form hydrogen bonds with water molecules. The lower boiling point of ethanethiol is explained by the lack of hydrogen bonding among molecules of this substance. On the other hand, hydrogen bonding occurs among ethanol molecules which results in a higher boiling point.

2. The products are $CH_2=CHCH_2OH$ and $(CH_3)_3COH$, respectively. Water is often the preferred nucleophile for introduction of OH when a stable and very reactive carbonium ion is involved, particularly where elimination is possible. Allyl alcohol is also made commercially by using Na_2CO_3-NaOH to establish a pH of 8-11.

3.
$$CH_3CH_2CH=CH_2 \text{ or } CH_3CH=CHCH_3 \xrightarrow{HOSO_2OH} CH_3CH_2\underset{OSO_2OH}{\overset{|}{C}}HCH_3$$

$$\xrightarrow{H_2O} CH_3CH_2\underset{OH}{\overset{|}{C}}HCH_3; \quad CH_3\underset{CH_3}{\overset{|}{C}}=CH_2 \xrightarrow{HOSO_2OH}$$

$$CH_3\underset{CH_3}{\overset{CH_3}{\overset{|}{\underset{|}{C}}}}-OSO_2OH \xrightarrow{H_2O} (CH_3)_3C-OH$$

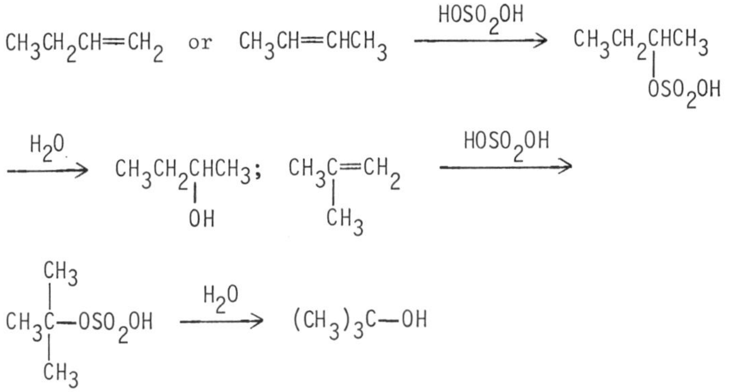

4.
$$CH_3CH=CH_2 \xrightarrow{H_2SO_4} CH_3\overset{\oplus}{C}HCH_3 \xrightarrow{H_2O} CH_3\underset{\oplus OH_2}{\overset{|}{C}}HCH_3 \xrightarrow[-H^{\oplus}]{} CH_3\underset{OH}{\overset{|}{C}}HCH_3$$

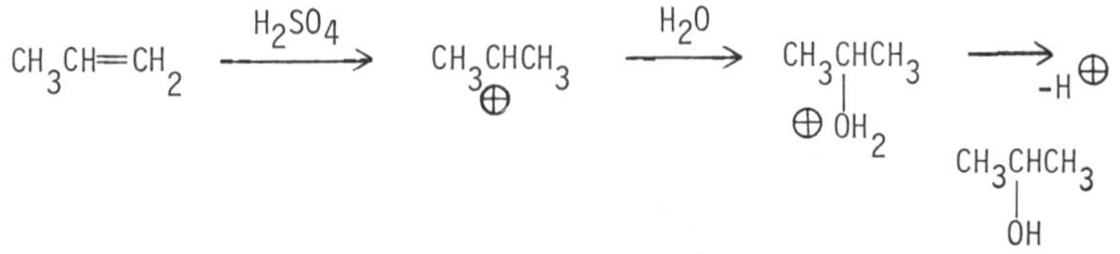

2° carbonium ion more stable than alternative 1° that could form.

5.

$$CH_3CH_2\underset{CH_2}{\overset{CH_3}{\underset{\|}{C}}} \xrightarrow{HOSO_2OH} CH_3CH_2\underset{CH_3}{\overset{CH_3}{\underset{|}{C}}}-OSO_2OH \xrightarrow{H_2O} CH_3CH_2\underset{CH_3}{\overset{CH_3}{\underset{|}{C}}}-OH$$

$$CH_3CH_2\overset{CH_3}{\underset{|}{C}}=CH_2 \xrightarrow[B_2H_6]{[BH_3]} (CH_3CH_2\overset{CH_3}{\underset{|}{C}}HCH_2\xrightarrow{})_3 B$$

"Markownikoff addition" gives 3° alcohol.

$$\xrightarrow{H_2O_2\text{-NaOH}} CH_3CH_2\overset{CH_3}{\underset{|}{C}}HCH_2OH$$

"Anti-Markownikoff addition" gives 1° alcohol.

6.
$$CH_3(CH_2)_5\overset{\delta\ominus}{C}H=\overset{\delta\oplus}{C}H_2 + H-OSO_2OH \xrightarrow{(1)} CH_3(CH_2)_5\underset{\oplus}{C}HCH_3 \xrightarrow{\ominus OSO_2OH} (2)$$

$$CH_3(CH_2)_5\overset{\delta\oplus}{C}HCH_3 \overset{\curvearrowleft}{\underset{OSO_2OH}{|}} \quad :\ddot{O}H_2 \xrightarrow{(3)} CH_3(CH_2)_5\underset{\oplus OH_2}{\overset{|}{C}HCH_3} \xrightarrow{HSO_4^{\ominus}} (4)$$

$$CH_3(CH_2)_5\underset{OH}{\overset{|}{C}HCH_3} + H_2SO_4$$

In step (1), addition of a proton to the terminal C gives a 2° carbonium ion; the alternative addition would give a less stable 1° carbonium ion. In step (2), the hydrogen sulfate anion adds to the carbonium ion center. In step (3), a water molecule (nucleophile) displaces the bisulfate anion. In step (4), the bisulfate anion assists in the removal of a proton. Alternatively, the 2° carbonium ion from step (1) can react with water and lose a proton to give the alcohol.

$$CH_3(CH_2)_5\overset{\delta\ominus}{C}H=CH_2 \xrightarrow[B_2H_6]{[BH_3]} [CH_3(CH_2)_7\xrightarrow{}]_3 B \xrightarrow{H_2O_2\text{-NaOH}} (2)$$
(1)

$$CH_3(CH_2)_7OH$$

In hydroboration, the boron atom is the acid (Lewis acid, because of the incomplete boron octet) rather than a proton and reacts with the terminal carbon. In step (2), alkaline hydrogen peroxide oxidation of octylborane gives the alcohol.

7. 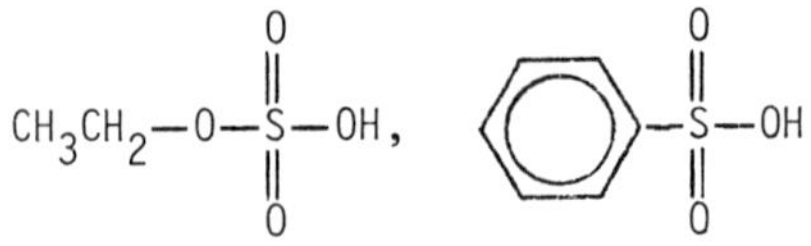 $CH_3CH_2-O-\overset{\overset{O}{\|}}{\underset{\underset{O}{\|}}{S}}-OH$, $C_6H_5-\overset{\overset{O}{\|}}{\underset{\underset{O}{\|}}{S}}-OH$

The hydrogen sulfate group contains four oxygen atoms, and there is a C—O—S sequence, while the sulfonic acid group has only three oxygen atoms and C attaches directly to S; C—S—OH.

8. $[(CH_3)_2CHO\!\!-\!]_4Ti$, $[(CH_3)_2CHO\!\!-\!]_2Ti$, $(CH_3)_2CHONa$

9.

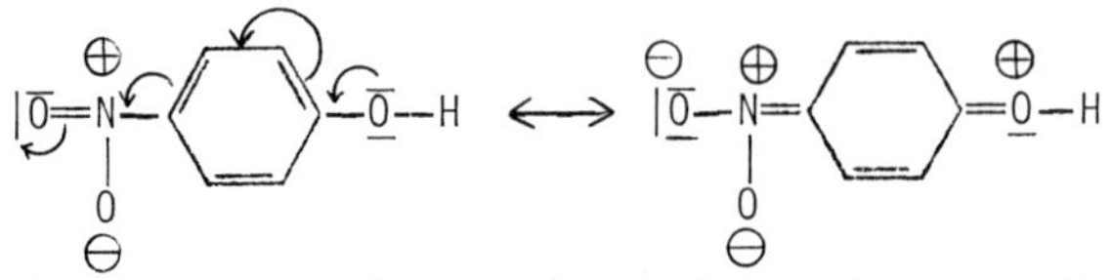

The NO_2 group enhances the positive charge on the OH oxygen and thus further weakens the OH bond.

10. 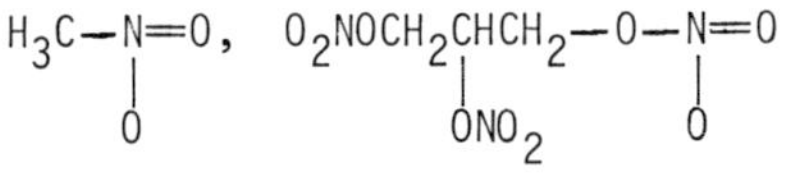 $H_3C-\underset{\underset{O}{|}}{N}=O$, $O_2NOCH_2\underset{\underset{ONO_2}{|}}{CH}CH_2-O-\underset{\underset{O}{|}}{N}=O$

In "nitroglycerin," the groups are actually nitrate groups, containing three oxygen atoms with a C—O—N sequence, while the nitro group contains only two oxygen atoms and direct attachment of C—N. "Nitroglycerin" is therefore a misnomer.

11. (1) $H_2C=CHCH_2Cl$ (2) $CH_3CH_2CH_2Cl$ (3) $(CH_3)_2CHBr$

(4)  cyclopentyl with CH_3 and I

12.

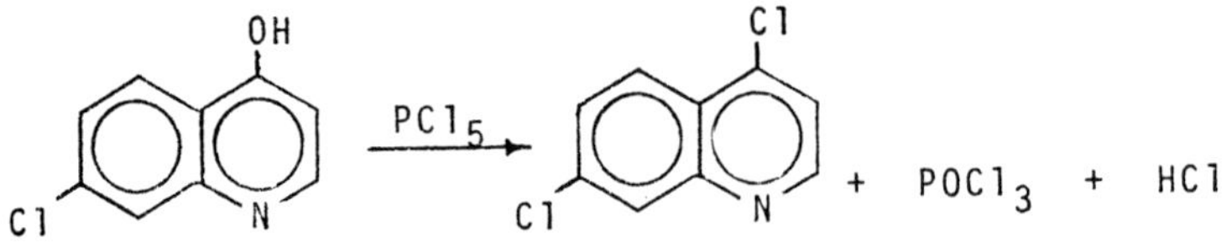

A hydroxyl group in the 3-position would not react, only in the 2- and 4-positions.

13.

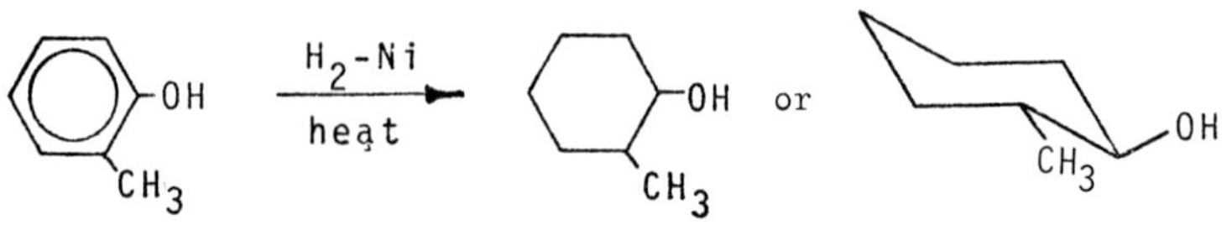

trans more stable, both groups equatorial. It is better to use the chair formula for the cyclohexane ring.

14. $CH_3(CH_2)_4-SH \xrightarrow[H_2O_2, NaOH]{[O]} CH_3(CH_2)_4-S-S-(CH_2)_4CH_3$

$\xleftarrow[Zn, H_2SO_4]{[H]}$

15. $(CH_3)_3C-OK + BrC_2H_5 \dashrightarrow (CH_3)_3C-O-C_2H_5$

    $(CH_3)_3C-Cl + NaOC_2H_5 \longrightarrow CH_3\underset{\underset{CH_3}{|}}{C}=CH_2 + C_2H_5OH + NaCl$

tert-Halides and, to a lesser extent, sec-halides undergo elimination with strong bases, so that the alternative scheme which would avoid this problem should be employed for the synthesis of unsymmetrical ethers.

16. 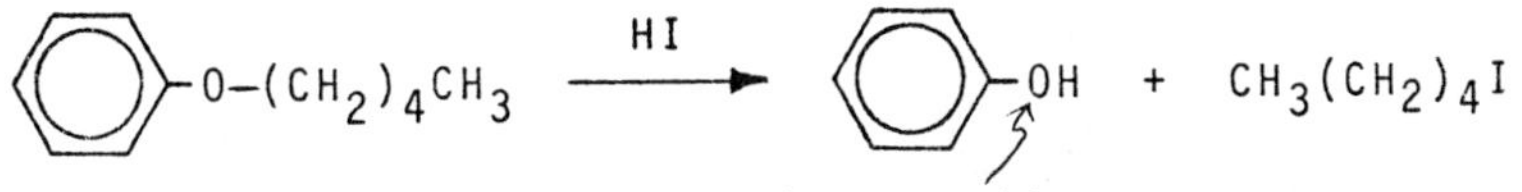 $C_6H_5-O-(CH_2)_4CH_3 \xrightarrow{HI} C_6H_5-OH + CH_3(CH_2)_4I$

Phenol cannot react with HI or other HX acids.

17. 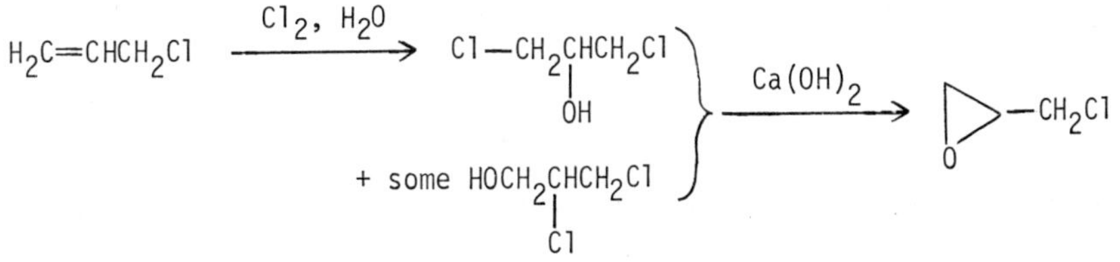 $H_2C=CHCH_2Cl \xrightarrow{Cl_2, H_2O} Cl-CH_2\underset{\underset{OH}{|}}{CH}CH_2Cl$

    $+$ some $HOCH_2\underset{\underset{Cl}{|}}{CH}CH_2Cl$

    $\xrightarrow{Ca(OH)_2}$ (epoxide)$-CH_2Cl$

18. 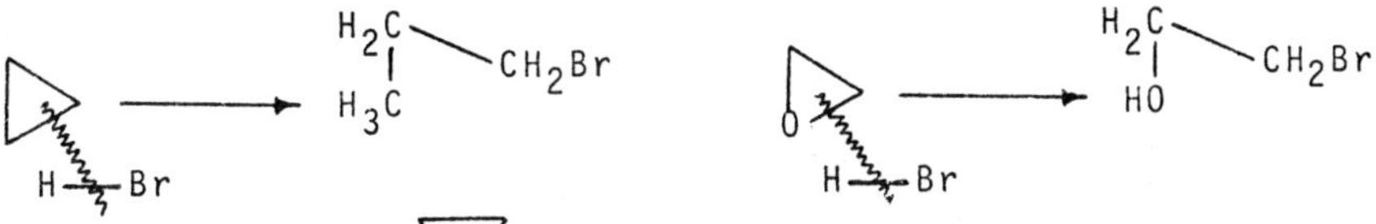 (cyclopropane) $\xrightarrow{H-Br}$ $H_2C(CH_2Br)(H_3C)$

    (cyclopropane with O) $\xrightarrow{H-Br}$ $H_2C(CH_2Br)(HO)$

19. 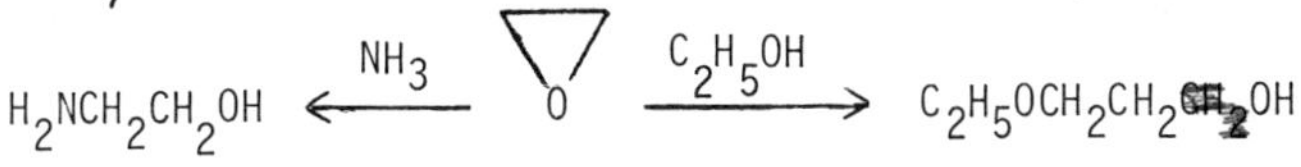 $H_2NCH_2CH_2OH \xleftarrow{NH_3}$ (epoxide) $\xrightarrow{C_2H_5OH} C_2H_5OCH_2CH_2OH$

20. (1) $C_6H_5-S-CH_3 \xrightarrow{HIO_4} C_6H_5-\underset{\underset{O}{\|}}{S}-CH_3 + HIO_3$

(2) 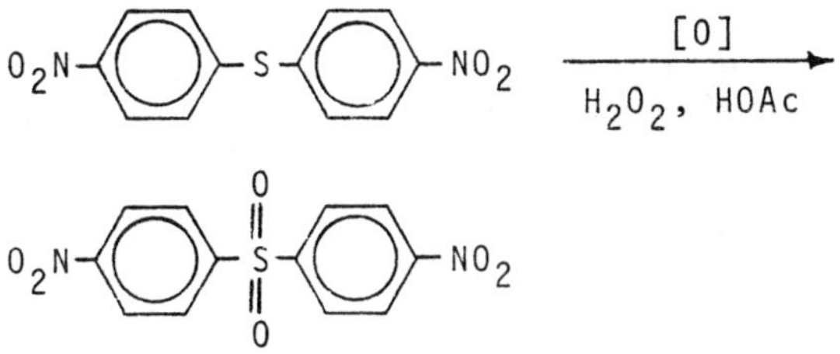 $O_2N-C_6H_4-S-C_6H_4-NO_2 \xrightarrow[H_2O_2, HOAc]{[O]}$

$O_2N-C_6H_4-\underset{\underset{O}{\overset{\overset{O}{\|}}{\|}}}{S}-C_6H_4-NO_2$

21.
(1) 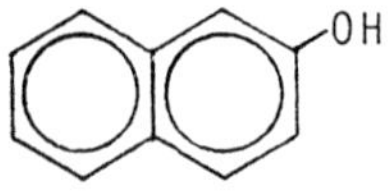 β-naphthol (2-naphthol) structure

(2) tert-butyl 4-methoxy-2-nitrophenyl ether

(3) $[(CH_3)_2CHO]_3Al$ (4) 2,5,5-trimethylhexanol

(5) 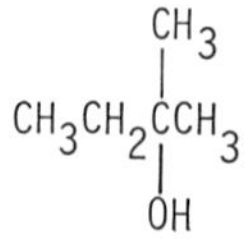 $CH_3CH_2\underset{\underset{OH}{|}}{\overset{\overset{CH_3}{|}}{C}}CH_3$ (6) allyl alcohol, 2-propen-1-ol

(7) $(CH_3)_3C-SH$ (8) isohexyl alcohol, 4-methyl-1-pentanol

(9) $[(CH_3)_2CH(CH_2)_3]_2O$ (10) benzyl sec-butyl sulfone

22.

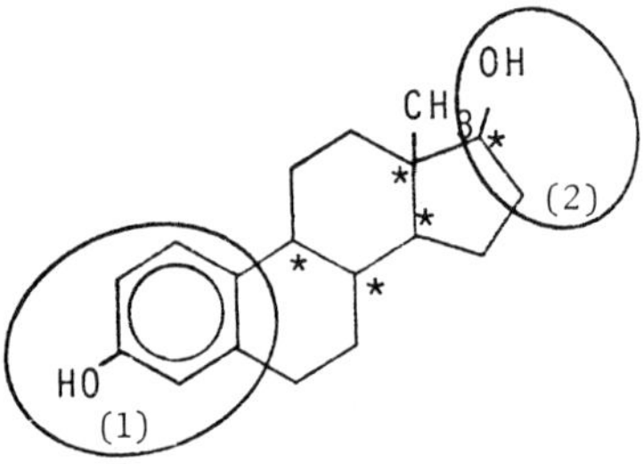

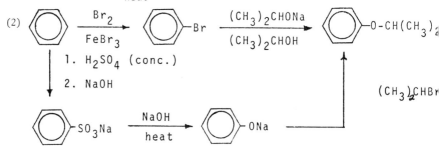

(1) phenolic hydroxyl group
(2) 2° hydroxyl group

2^5 = 32 optical isomers
16 racemic modifications

23.

(1) $H_2C=CH-CH_3 \xrightarrow[\text{heat}]{Cl_2} H_2C=CHCH_2Cl \xrightarrow{OH^\ominus} H_2C=CHCH_2OH$

(2) $C_6H_6 \xrightarrow[FeBr_3]{Br_2} C_6H_5-Br \xrightarrow[(CH_3)_2CHOH]{(CH_3)_2CHONa} C_6H_5-O-CH(CH_3)_2$

1. H_2SO_4 (conc.)
2. NaOH

$C_6H_5-SO_3Na \xrightarrow[\text{heat}]{NaOH} C_6H_5-ONa$

$(CH_3)_2CHBr$

In the first sequence, treatment of bromobenzene with an alkoxide would give a benzyne intermediate.

(3) 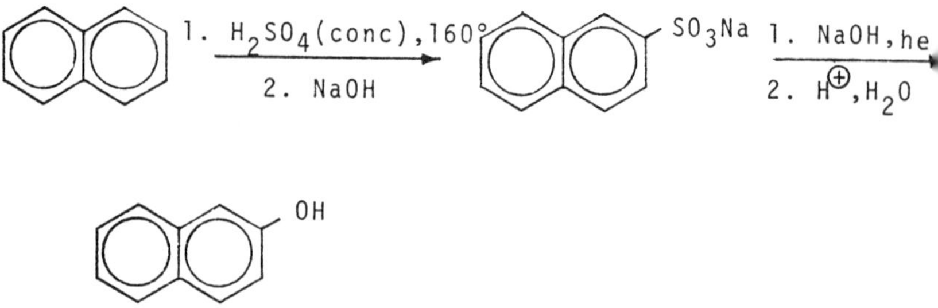 naphthalene $\xrightarrow[\text{2. NaOH}]{\text{1. }H_2SO_4(conc), 160°}$ naphthalene-SO$_3$Na $\xrightarrow[\text{2. }H^\oplus, H_2O]{\text{1. NaOH, heat}}$

2-naphthol structure

At 80°, 1-naphthalenesulfonic acid is the chief product, and 1-naphthol can be derived from it by a similar route.

(4) $CH_3CH=CH_2 \xrightarrow{Cl_2, H_2O} CH_3\underset{OH}{CHCH_2Cl} \xrightarrow{Ca(OH)_2}$ CH_3-cyclopropane-oxide (methyloxirane)

$\xrightarrow{H_2O} CH_3\underset{OH}{CHCH_2OH}$

Direct oxidation of the starting material with $KMnO_4$ solution will give the same product but probably not in as good yield.

24.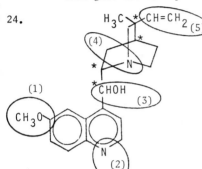

(1) methyl phenyl ether group
(2) aromatic, heterocyclic <u>tert</u> amino group; quinoline <u>ring</u> system
(3) 2° hydroxyl group
(4) heterocyclic, bridge-head <u>tert-</u> amino group
(5) vinyl double bond

2^4 = 16 optical isomers
    8 racemic modifications

(1) and (3) are particularly relevant to this chapter.

25. [structure: cyclopentadiene with Cl, Cl, Cl, Cl, OCH_3, OCH_3]

Two ether groups are introduced by substitution of the very reactive allyl-type chloro groups. The relatively unreactive vinyl-type chloro groups are not displaced.

26. $(C_2H_5O)_2\underset{\underset{S}{\|}}{P}-Cl + NaO-\langle\bigcirc\rangle-NO_2 \longrightarrow$

$(C_2H_5O)_2\underset{\underset{S}{\|}}{P}-O-\langle\bigcirc\rangle-NO_2$

parathion

27. [structure with CH_3, OH (2), H_3C, H_3C, O (1), (CH_2)_4CH_3]

(1) <u>tert</u>-alkyl aryl heterocyclic ether
(2) phenolic hydroxyl group
(3) carbon-carbon double bond

2^2 = 4 optical isomers
    2 racemic modifications

AMINES, DYES, AND ALKALOIDS

1. Compared with methylamine, methanol and methyl fluoride are relatively nonbasic, because the nonbonding electron pairs on oxygen and fluorine are held more tightly to the nucleus by the greater positive nuclear charge. Thus, they are less available for bonding than the one nonbonding electron pair on nitrogen. Even so, alcohols and ethers are basic in concentrated mineral acid.

$$CH_3-\underline{\overline{O}}-H + H_2SO_4 \longrightarrow CH_3-\overset{\oplus}{\underset{H}{\overline{O}}}-H \; \; HSO_4^{\ominus}$$

   Trivalent oxygen is, however, unstable so that stable oxonium salts cannot ordinarily be isolated.

2. Logically, while meta-directing groups, situated <u>ortho</u> or <u>para</u>, increase acidity of phenols, they decrease basicity of aromatic amines. In comparison with Pb. 10.9:

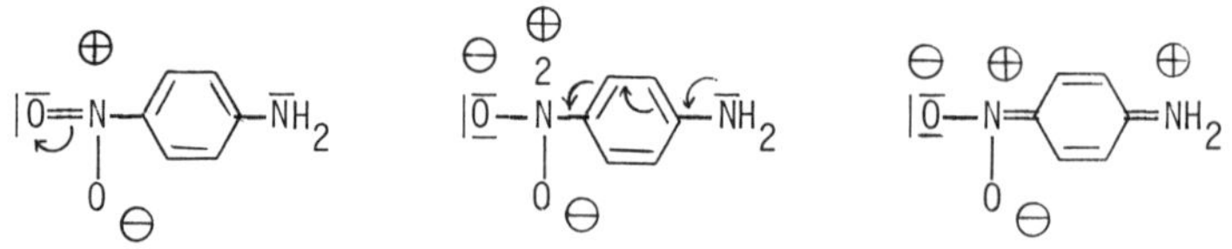

   Development of a positive charge on nitrogen makes it less basic to the extent that the nonbonding electron pair is less available for bonding with a proton. The same reasoning relative to oxygen in the phenols makes easier release of a proton understandable.

3. (1) $ClCH_2CH_2Cl \xrightarrow[\text{2. NaOH}]{2NH_3, \; 150°} H_2NCH_2CH_2NH_2$

   (2) $CH_3CH_2Cl + H_2NCH_2CH_2CH_3 \xrightarrow[\text{2. NaOH}]{} CH_3CH_2NHCH_2CH_2CH_3$

   (3)  (epoxide) $\xrightarrow{NH_3} HOCH_2CH_2NH_2$

(4) C₆H₅-NHCH₃ + CH₃I ⟶ C₆H₅-N(CH₃)₂

(5) H₃N + ClCH₂CO₂H ⟶ H₂NCH₂CO₂H
                                      glycine

4. 1,3-dinitrobenzene (with additional NO_2) $\xrightarrow[NH_4SH]{[H]}$ 3-nitroaniline (NO_2, NH_2)

5. methylcyclohexane (CH_3) + [NCl₃] $\xrightarrow[NH_4Cl-HCl]{1.\ Ca(OCl)_2}$ 1-methyl-1-aminocyclohexane (H_3C, NH_2)

                           $AlCl_3$, CH_2Cl_2
                           2. H_2O

6. $C_6H_5SO_2Cl + \begin{cases} H_2NC_6H_5 \\ CH_3NHC_6H_5 \\ (CH_3)_2NC_6H_5 \end{cases} \xrightarrow{NaOH} C_6H_5SO_2\overset{\ominus}{N}C_6H_5\ \overset{\oplus}{Na} + \begin{bmatrix} C_6H_5SO_2\underset{CH_3}{N}C_6H_5 \\ + \\ (CH_3)_2NC_6H_5 \end{bmatrix}$

                                                  (1)   (2)

Compound (1), derived from aniline, has an acidic hydrogen, is thus soluble in the aqueous sodium hydroxide, and can be separated by filtration from the mixture (2) of the nonacidic sulfonamide and the unreacted dimethylaniline (the latter is unable to react with dimethylaniline because there is no replaceable hydrogen). Mixture (2) may be separated, as the basic N,N-dimethylaniline can be extracted by hydrochloric acid from the neutral N-methyl-N-phenyl-benzenesulfonamide. Once separated, the amine hydrochloride is neutralized to give the free amine.

$C_6H_5N(CH_3)_2 \cdot HCl \xrightarrow{NaOH} C_6H_5N(CH_3)_2 + NaCl + H_2O$

The sulfonamides can be hydrolyzed with strong acid to regenerate the amines.

7. HO–C₆H₄–NH₂ $\xrightarrow[2.\ NaOAc]{1.\ Ac_2O}$ HO–C₆H₄–NHCOCH₃   p-acetamidophenol (Tylenol)

8. The reaction sequence leading to benzenediazonium cation from aniline is identical with the sequence leading to the methyldiazonium cation in 11.17. However, the benzenediazonium cation is relatively stable because of delocalization of the positive charge over the benzene ring.

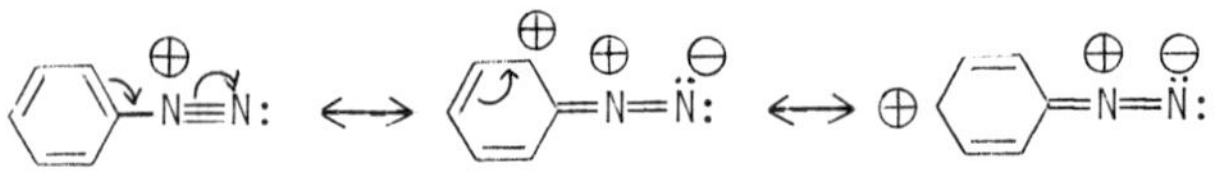

9. 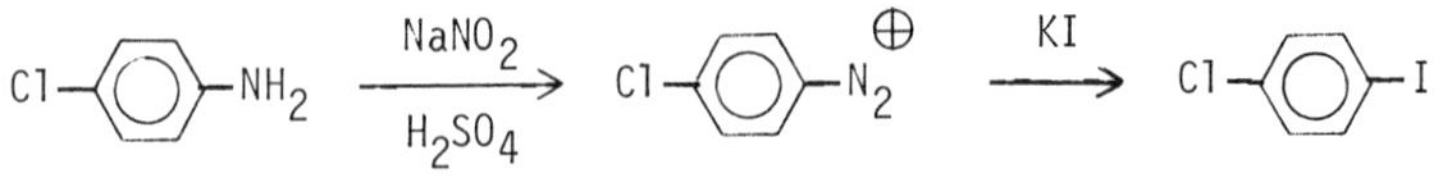
NaNO2
H2SO4
KI

10. 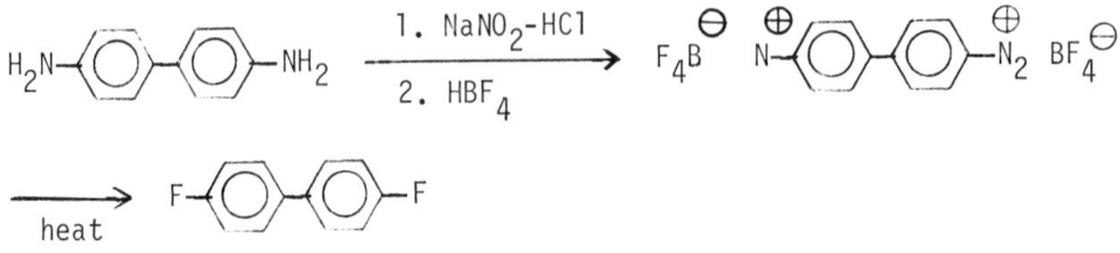
1. NaNO2-HCl
2. HBF4
heat

See answer to Pb. 20 for the source of the starting material.

11. 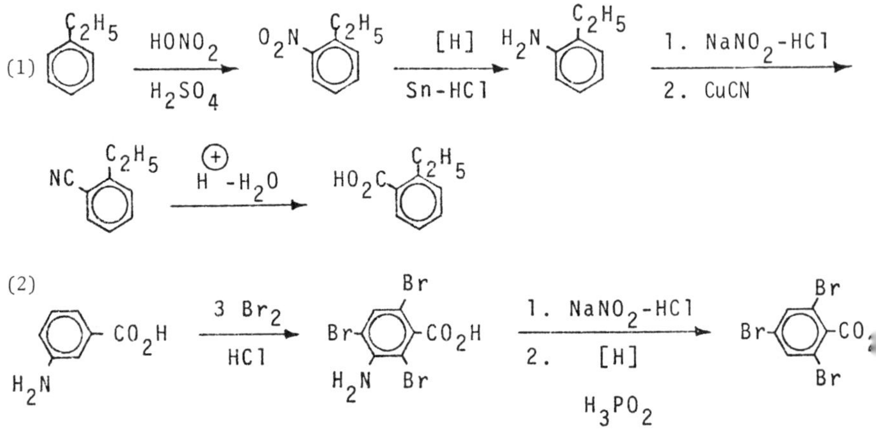
(1)
HONO2
H2SO4
[H]
Sn-HCl
1. NaNO2-HCl
2. CuCN
H⊕-H2O
(2)
3 Br2
HCl
1. NaNO2-HCl
2. [H]
H3PO2

12. 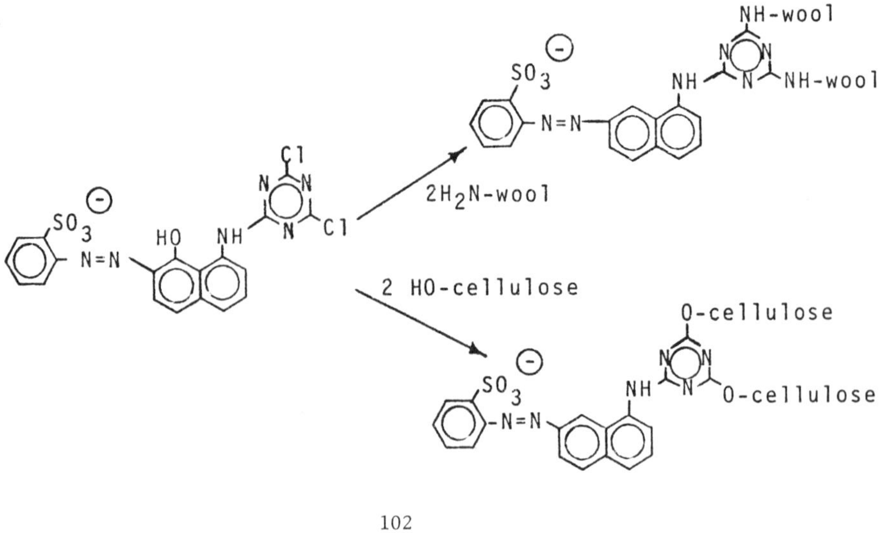
2H2N-wool
2 HO-cellulose
NH-wool
NH-wool
O-cellulose
O-cellulose

13. A water-soluble reduced form of indigo (indigo white) is obtained by reduction with alkaline sodium hydrosulfite ($Na_2S_2O_4$). (Prior to the modern era of chemical industry, the reduction was a fermentation process.) The textile is then immersed in the indigo white. Exposure to air converts indigo white to indigo by air oxidation. The fiber is thus impregnated with the insoluble blue dye.

14.

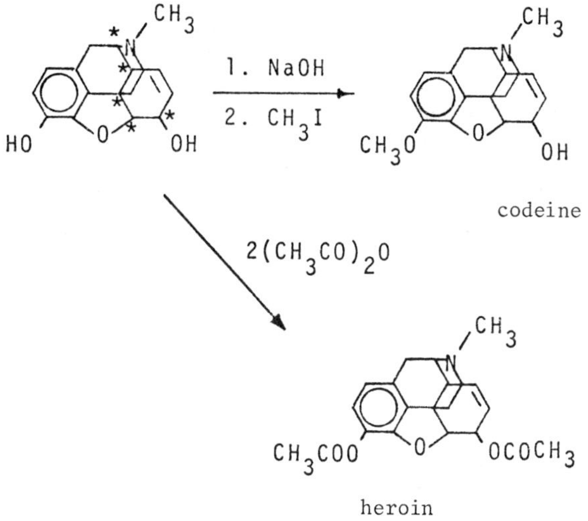

$2^5 = 32$ optical isomers
16 racemic modifications

codeine

$2(CH_3CO)_2O$

heroin

The structures of codeine and heroin should be remembered on the basis of the morphine structure as a reference.

15.

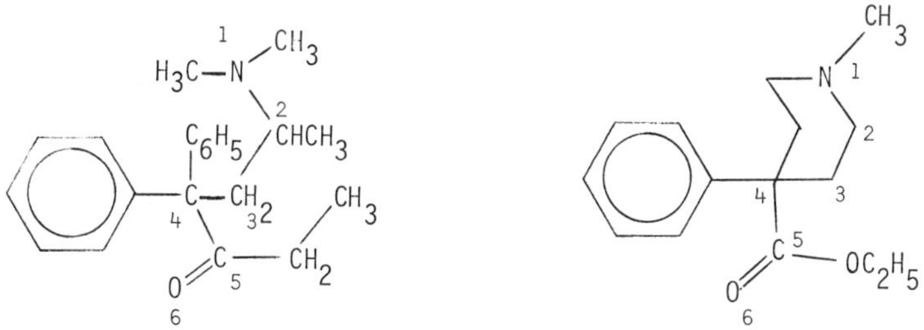

The position of the nitrogen and oxygen atoms and the phenyl group are identical with those in morphine, as numbering shows. This is just one consideration of a number studied in the synthesis of compounds of medicinal interest.

16. (1) tripropylamine (2) $(CH_3)_2CHCH_2NHC_2H_5$
    (3) $(CH_3)_3CNH_2$ (4) N-methyl-N-propylpentylamine
    (5) m-phenylenediamine
    (6) 6-dimethylamino-4,4-diphenyl-3-heptanone

(7) anthracen-1-yl-NHCH$_2$CH$_2$CH$_3$ (1-(propylamino)anthracene)

(8) Cl—C$_6$H$_4$—N$_2^{\oplus}$Cl$^{\ominus}$

(9) 5-amino-2-chlorophenol; OH has priority over an NH$_2$ group

(10) 5-amino-2-heptanol

(11) cyclopentyl—N(CH$_3$)$_3^{\oplus}$ Cl$^{\ominus}$

(12) $\underline{N}$-ethyl-1,3-dimethyl-5-phenyl-$\underline{N}$-propylhexylamine

17.

$$C_6H_5CH_2\underset{CH_3}{\overset{|}{C}H}-NH_2 \xrightarrow[NaOH]{C_6H_5SO_2Cl} C_6H_5CH_2\underset{CH_3}{\overset{|}{C}H}-\overset{\ominus}{N}SO_2C_6H_5 \;\; Na^{\oplus} \quad \text{soluble}$$

$$C_6H_5CH_2\underset{CH_3}{\overset{|}{C}H}-\underset{CH_3}{\overset{|}{N}H} \xrightarrow{C_6H_5SO_2Cl} C_6H_5CH_2\underset{CH_3}{\overset{|}{C}H}-\underset{CH_3}{\overset{|}{N}}-SO_2C_6H_5 \quad \text{insoluble}$$

$$C_6H_5CH_2\underset{CH_3}{\overset{|}{C}H}-NH_2 \xrightarrow[NaNO_2, HCl]{[HONO]} C_6H_5CH_2\underset{CH_3}{\overset{|}{C}H^{\oplus}} + N_2$$

$$C_6H_5CH_2\overset{\oplus}{C}HCH_2CH_3 \quad \xrightarrow{\text{more stable}} \quad \begin{cases} \xrightarrow{Cl^-} C_6H_5\underset{Cl}{\overset{|}{C}H}CH_2CH_3 \\ \xrightarrow{H_2O} C_6H_5\underset{OH}{\overset{|}{C}H}CH_2CH_3 \end{cases}$$

$$C_6H_5CH=CHCH_3 \xleftarrow{-H^{\oplus}} C_6H_5\overset{\oplus}{C}HCH_2CH_3$$

Positive carbon next to benzene ring is more stable because of delocalization with ring; therefore, rearrangement occurs.

$$C_6H_5CH_2\underset{CH_3}{\overset{|}{C}H}-\underset{CH_3}{\overset{|}{N}H} \xrightarrow[NaNO_2, HCl]{[HONO]} C_6H_5CH_2\underset{CH_3}{\overset{|}{C}H}-\underset{CH_3}{\overset{|}{N}}-NO$$

18.

(1) 2-aminoanthracene $\xrightarrow{(CH_3CO)_2O}$ 2-(N-acetylamino)anthracene (Ar—NHCOCH$_3$)

(2) 3,4,5-trimethoxy-C$_6$H$_2$—CH$_2$CH$_2$NH$_2$ $\xrightarrow{HI (XS)}$ 3,4,5-trihydroxy-C$_6$H$_2$—CH$_2$CH$_2$NH$_2 \cdot$HI + 3CH$_3$I

Two reactions occur: (1) cleavage of the ether groups with hydriodic acid and (2) formation of an ammonium salt. In acidic solution, methylation of the amino group by methyl iodide does not take place.

(3) 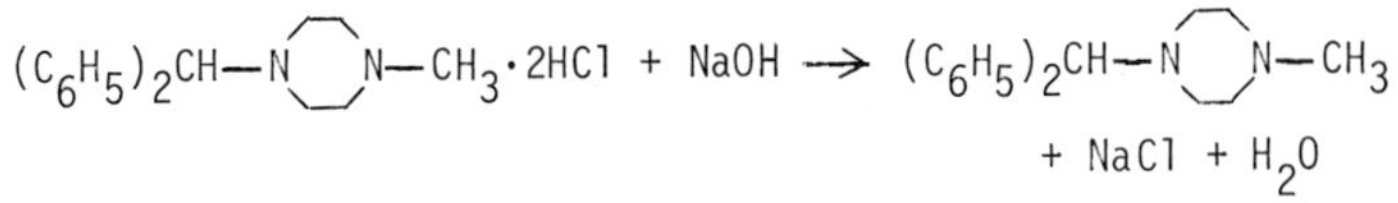 $(C_6H_5)_2CH-N\overset{\frown}{}N-CH_3 \cdot 2HCl + NaOH \longrightarrow (C_6H_5)_2CH-N\overset{\frown}{}N-CH_3 + NaCl + H_2O$

(4) $C_2H_5-C_6H_4-NH_2 \xrightarrow{2Cl_2} C_2H_5-C_6H_2(Cl)_2-NH_2$

(5) $C_2H_5-C_6H_4-NHCOCH_3 \xrightarrow{Cl_2} C_2H_5-C_6H_3(Cl)-NHCOCH_3$

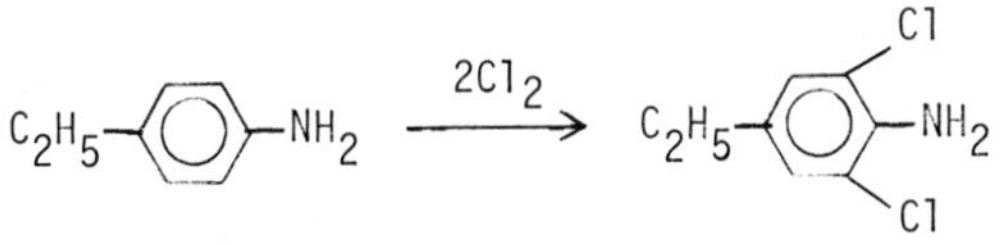

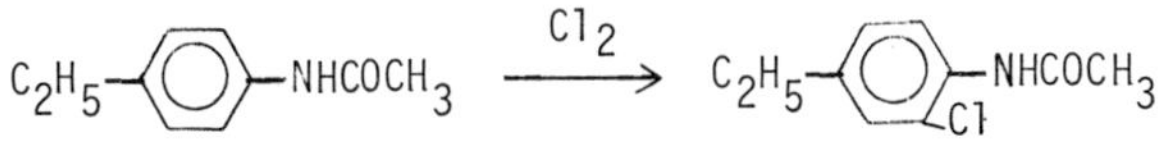

(6) 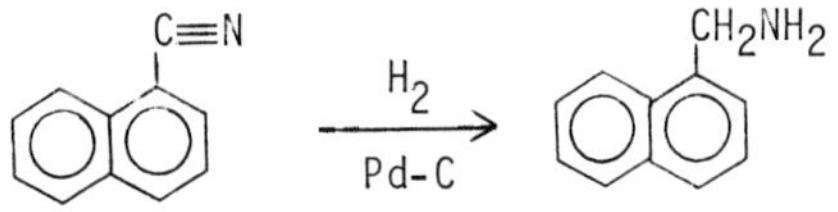 naphthalene-C≡N $\xrightarrow{H_2, Pd-C}$ naphthalene-CH$_2$NH$_2$

(7) 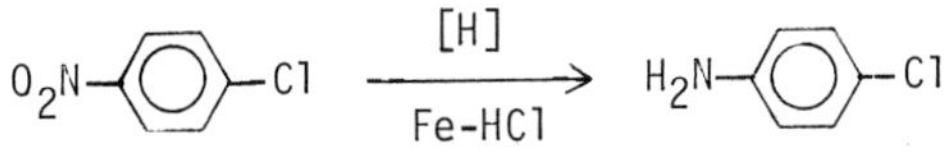 $O_2N-C_6H_4-Cl \xrightarrow[Fe-HCl]{[H]} H_2N-C_6H_4-Cl$

(8) $H_2N-(CH_2)_6-NH_2 + CH_3Cl \text{ (xs)} \xrightarrow{\text{heat}} (CH_3)_3\overset{\oplus}{N}-(CH_2)_6-\overset{\oplus}{N}(CH_3)_3 \; 2Cl^{\ominus}$

(9) 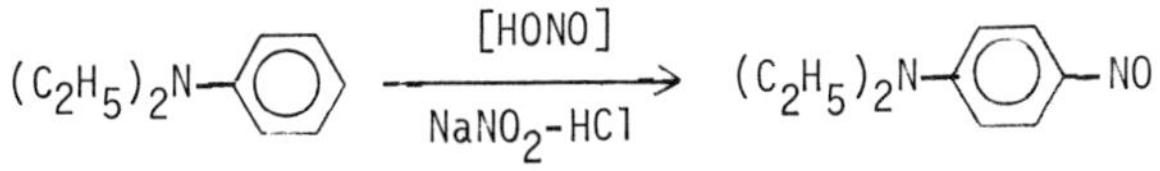 $(C_2H_5)_2N-C_6H_5 \xrightarrow[NaNO_2-HCl]{[HONO]} (C_2H_5)_2N-C_6H_4-NO$

(10) $HOCH_2CH_2NH_2 \xrightarrow[NaOH]{C_6H_5COCl} HOCH_2CH_2\overset{H}{N}-\underset{O}{\overset{\parallel}{C}}C_6H_5$

19. (1) Only the aromatic amine will react with nitrous acid to form a stable diazonium salt, which will couple with 2-naphthol to give an orange dye:

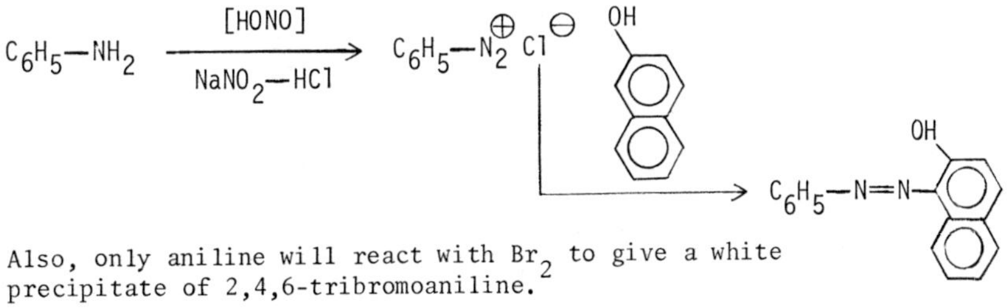

$C_6H_5-NH_2 \xrightarrow[NaNO_2-HCl]{[HONO]} C_6H_5-\overset{\oplus}{N_2} \; Cl^{\ominus}$ + 2-naphthol (OH) $\longrightarrow C_6H_5-N=N-$(naphthol-OH)

Also, only aniline will react with Br_2 to give a white precipitate of 2,4,6-tribromoaniline.

(2) Only p-nitroaniline is basic enough to react with hydrochloric acid to give a clear solution of amine salt. The 2,4,6-trinitroaniline is as nonbasic as an amide and is soluble in water or HCl.

$$O_2N-C_6H_4-NH_2 \xrightarrow{HCl} O_2N-C_6H_4-NH_2 \cdot HCl$$

(3) With benzenesulfonyl chloride under alkaline conditions propylamine would give a clear solution and dipropylamine would give a white precipitate:

$$CH_3CH_2CH_2NH_2 \xrightarrow{C_6H_5SO_2Cl} CH_3CH_2CH_2-\overset{\ominus}{N}-SO_2C_6H_5 \;\; \overset{\oplus}{Na}$$
clear solution

$$(CH_3CH_2CH_2)_2NH \xrightarrow[NaOH]{C_6H_5SO_2Cl} (CH_3CH_2CH_2)_2NSO_2C_6H_5 \downarrow$$
white precipitate

Nitrous acid could also be used. Propylamine would react to evolve nitrogen, while dipropylamine would give a yellow precipitate.

(4) p-Aminoacetophenone has a basic amino group and would thus dissolve in hydrochloric acid, while acetanilide is a nonbasic amide and would not dissolve:

$$CH_3CO-C_6H_4-NH_2 \xrightarrow{HCl} CH_3CO-C_6H_4-NH_2 \cdot HCl$$
clear solution

(5) With nitrous acid, N-ethylaniline gives a yellow N-nitroso derivative (Caution: carcinogenic properties); N,N-diethylaniline gives a green p-C-nitroso derivative. (Benzenesulfonyl chloride could also be employed as a test.)

$$C_2H_5NH-C_6H_5 \xrightarrow[NaNO_2-HCl]{[HONO]} C_2H_5\overset{|}{N}-C_6H_5$$
$$\qquad\qquad\qquad\qquad\qquad\qquad NO$$
yellow

$$(C_2H_5)_2N-C_6H_5 \xrightarrow[NaNO_2-HCl]{[HONO]} (C_2H_5)_2N-C_6H_4-NO$$
green

(6) Benzanilide is a neutral amide, insoluble in sodium hydroxide solution, while benzenesulfonamide is acidic and will dissolve in aqueous alkali; see part (3) above.

20.

$\text{C}_6\text{H}_5\text{-NO}_2 \xrightarrow[\text{[H]}]{\text{Zn-NaOH}} \text{C}_6\text{H}_5\text{-NHNH-C}_6\text{H}_5 \xrightarrow{\text{HCl}} \text{H}_2\text{N-C}_6\text{H}_4\text{-C}_6\text{H}_4\text{-NH}_2$

hydrazobenzene (2)   benzidine (3)

The transformation of (2) to (3) is called the benzidine rearrangement. The product (3) is important in the manufacture of certain dyes.

21.

(1) $\text{Cl-C}_6\text{H}_4\text{-NO}_2 \xrightarrow{\text{CH}_3\text{SNa}} \text{CH}_3\text{S-C}_6\text{H}_4\text{-NO}_2 \xrightarrow[\text{Sn-HCl}]{\text{[H]}}$

$\text{CH}_3\text{S-C}_6\text{H}_4\text{-NH}_2 \xrightarrow{\text{C}_6\text{H}_5\text{COCl}} \text{CH}_3\text{S-C}_6\text{H}_4\text{-NHCOC}_6\text{H}_5$

(2) $\text{CH}_2\text{=CH}_2 \xrightarrow{\text{Br}_2} \text{BrCH}_2\text{CH}_2\text{Br} \xrightarrow{\text{NaCN}} \text{N≡CCH}_2\text{CH}_2\text{C≡N} \xrightarrow[\text{Pd-C}]{\text{H}_2}$

$\text{H}_2\text{N-(CH}_2)_4\text{-NH}_2$

(3) 1-naphthylamine $\xrightarrow[\text{2. HBF}_4]{\text{1. NaNO}_2\text{-HCl}}$ 1-naphthyl-$\text{N}_2^{\oplus}\text{BF}_4^{\ominus}$ $\xrightarrow{\text{heat}}$ 1-fluoronaphthalene + N_2 + BF_3

(4) $\text{C}_2\text{H}_5\text{-C}_6\text{H}_5 \xrightarrow{\text{HONO}_2\text{-H}_2\text{SO}_4} \text{C}_2\text{H}_5\text{-C}_6\text{H}_4\text{-NO}_2 \xrightarrow[\text{Sn-HCl}]{\text{[H]}}$

$\text{C}_2\text{H}_5\text{-C}_6\text{H}_4\text{-NH}_2 \xrightarrow[\text{2. H}_2\text{O, heat}]{\text{1. NaNO}_2\text{-H}_2\text{SO}_4} \text{C}_2\text{H}_5\text{-C}_6\text{H}_4\text{-OH}$

(5) $\text{C}_6\text{H}_5\text{NO}_2 \xrightarrow[\text{FeBr}_3]{\text{Br}_2} m\text{-Br-C}_6\text{H}_4\text{-NO}_2 \xrightarrow[\text{Sn-HCl}]{\text{[H]}} m\text{-Br-C}_6\text{H}_4\text{-NH}_2$

(6) 3-aminopyridine $\xrightarrow[\text{2. CuBr}]{\text{1. NaNO}_2\text{-H}_2\text{SO}_4}$ 3-bromopyridine

(7) $\text{H}_2\text{N-C}_6\text{H}_4\text{-C}_6\text{H}_4\text{-NH}_2 \xrightarrow[\text{2. CuCN}]{\text{1. NaNO}_2\text{-H}_2\text{SO}_4} \text{NC-C}_6\text{H}_4\text{-C}_6\text{H}_4\text{-CN}$

$\xrightarrow[\text{H}_2\text{SO}_4, \text{ heat}]{\text{H}_2\text{O}} \text{HO}_2\text{C-C}_6\text{H}_4\text{-C}_6\text{H}_4\text{-CO}_2\text{H}$

(8) $p\text{-CH}_3\text{-C}_6\text{H}_4\text{-NH}_2 \xrightarrow{\text{Ac}_2\text{O}} p\text{-CH}_3\text{-C}_6\text{H}_4\text{-NHAc} \xrightarrow{\text{HNO}_3\text{-H}_2\text{SO}_4}$ (CH$_3$, NHAc, NO$_2$ substituted benzene) $\xrightarrow[\text{heat}]{\text{H}^{\oplus}\text{-H}_2\text{O}}$

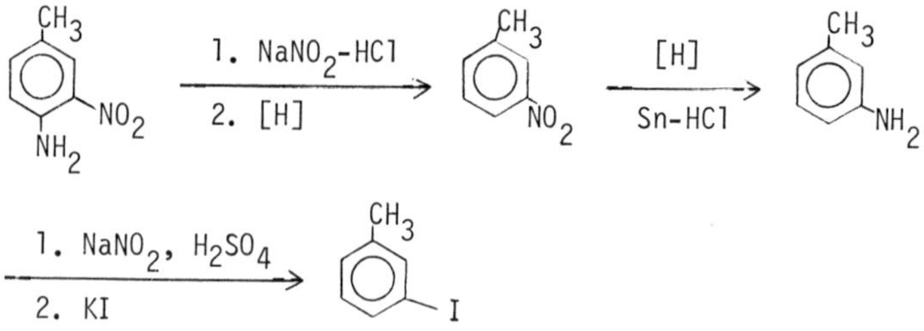

$$\underset{\underset{NH_2}{|}}{\overset{\overset{CH_3}{|}}{\bigcirc}}\!\!-NO_2 \xrightarrow[2.\ [H]]{1.\ NaNO_2-HCl} \underset{NO_2}{\overset{CH_3}{\bigcirc}} \xrightarrow[Sn-HCl]{[H]} \underset{NH_2}{\overset{CH_3}{\bigcirc}}$$

$$\xrightarrow[2.\ KI]{1.\ NaNO_2,\ H_2SO_4} \underset{I}{\overset{CH_3}{\bigcirc}}$$

22.

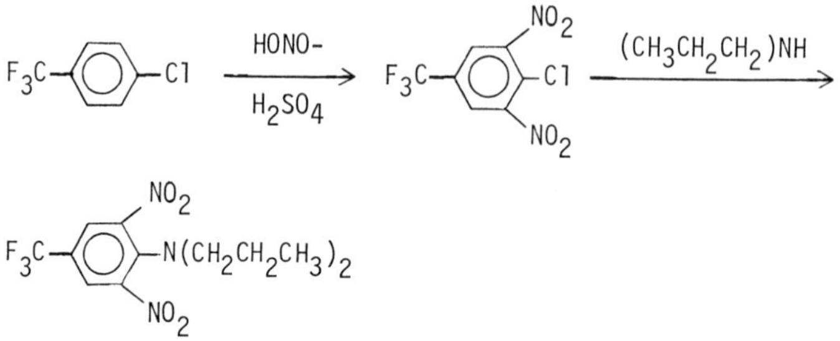

$$F_3C-\bigcirc-Cl \xrightarrow[H_2SO_4]{HONO-} F_3C-\underset{NO_2}{\overset{NO_2}{\bigcirc}}-Cl \xrightarrow{(CH_3CH_2CH_2)NH}$$

$$F_3C-\underset{NO_2}{\overset{NO_2}{\bigcirc}}-N(CH_2CH_2CH_3)_2$$

23.

(1) 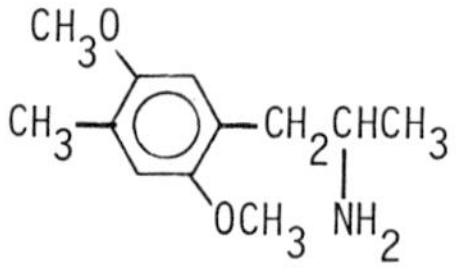 $\underset{OCH_3}{\overset{CH_3O}{CH_3-\bigcirc-CH_2\underset{NH_2}{\overset{|}{C}}HCH_3}}$

Note the similarity in structure to benzedrine and mescaline. STP has no basis in chemical nomenclature but is rather a term from the drug culture.

(2) 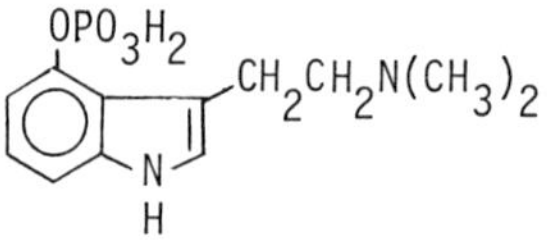 $\overset{OPO_3H_2}{\underset{\underset{H}{N}}{\bigcirc\!\!\!\Vert}}-CH_2CH_2N(CH_3)_2$

ALDEHYDES AND KETONES

1.

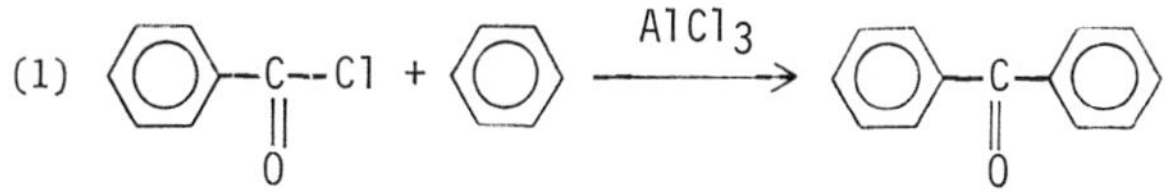

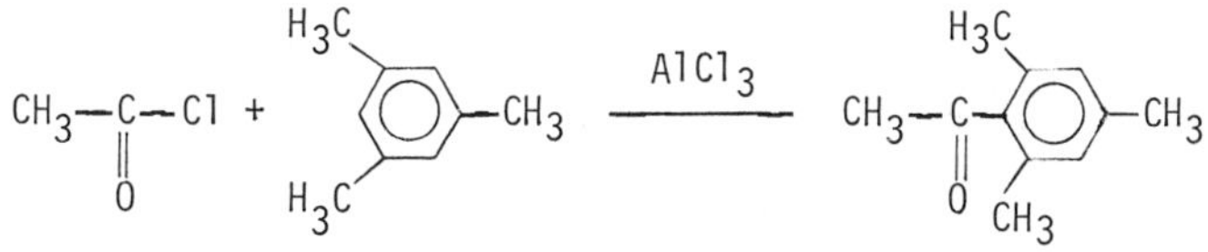

(1) $C_6H_5{-}\underset{\underset{O}{\|}}{C}{-}Cl + C_6H_6 \xrightarrow{AlCl_3} C_6H_5{-}\underset{\underset{O}{\|}}{C}{-}C_6H_5$

(2) $CH_3{-}\underset{\underset{O}{\|}}{C}{-}Cl + \text{2,4,6-(CH}_3\text{)}_3\text{C}_6\text{H}_2\text{H} \xrightarrow{AlCl_3} CH_3{-}\underset{\underset{O}{\|}}{C}{-}\text{C}_6\text{H}_2\text{(CH}_3\text{)}_3$

2. $\text{cyclo-C}_5\text{H}_9{-}\underset{\underset{O}{\|}}{C}{-}OH + CH_3Li \longrightarrow \text{cyclo-C}_5\text{H}_9{-}\underset{\underset{O}{\|}}{C}{-}CH_3 + LiOH$

3. $CH_4 \xrightarrow[\text{heat}]{O_2} CO + H_2 \xrightarrow{\text{ZnO-Cr}_2\text{O}_3, 350°, \text{pressure}} CH_3OH \xrightarrow[600°]{Ag}$

$HCHO + H_2\uparrow$
formaldehyde

$CH_3{-}CH_3 \xrightarrow[\text{pressure}]{800°} H_2C{=}CH_2 \xrightarrow{1.\ H_2SO_4;\ 2.\ H_2O} CH_3CH_2OH \xrightarrow[275°]{Cu}$

$CH_3CHO + H_2\uparrow$
acetaldehyde

$\text{cracking} \longrightarrow CH_3CH{=}CH_2 \xrightarrow{1.\ H_2SO_4;\ 2.\ H_2O} CH_3{-}\underset{\underset{OH}{|}}{CH}{-}CH_3 \xrightarrow[300°]{CuO}$

$CH_3{-}\underset{\underset{O}{\|}}{C}{-}CH_3$
acetone

4. $CH_3(CH_2)_2CHO + 2[CuO] \xrightarrow[NaOH]{Cu^{2+}\text{-tartrate}} CH_3(CH_2)_2CO_2Na + Cu_2O\downarrow$

5.

<img of cyclohexanone with δ+ C=O attacked by δ- NaHSO_3 → cyclohexane with OH and SO_3Na>

Note that the sulfur atom of the sulfonate group attaches to the carbon of the carbonyl group.

6. $C_6H_5CH_2CH_2MgBr + \triangle O \xrightarrow[2.\ H^+-H_2O]{1.\ \text{dry Et}_2O} C_6H_5CH_2CH_2CH_2CH_2OH$

7. $CH_3CH_2CH_2MgBr + C_6H_5CCH_2CH(CH_3)_2 \atop \|\atop O$

$C_6H_5MgBr + CH_3CH_2CH_2CCH_2CH(CH_3)_2 \atop \|\atop O \longrightarrow CH_3CH_2CH_2\underset{\underset{OH}{|}}{\overset{\overset{C_6H_5}{|}}{C}}CH_2CH(CH_3)_2$

$(CH_3)_2CHCH_2MgBr + C_6H_5CCH_2CH_2CH_3 \atop \|\atop O$

After acidification.

8.

<cyclohexanone> $+ HOCH_2CH_2OH \xrightarrow{\underline{p}-CH_3C_6H_4SO_3H}$ <cyclic acetal product>

One should always be alert to interaction of functional groups which would lead to 5- or 6-membered rings. In this case, a cyclic acetal is formed because both of the OH groups required come from the same molecule. Although acetals of ketones (ketals) do not usually form by the direct method, they do when a cyclic product is possible.

9. $F\text{-}\bigcirc\text{-}\underset{H}{C}=O + H_2NNHC_6H_5 \longrightarrow F\text{-}\bigcirc\text{-}\underset{H}{C}=NNHC_6H_5 + H_2O$

Instead of going through the details of addition-elimination in each instance, the reaction may be viewed as a "condensation" with a direct elimination of water between the two reactants. The reactions involve the same type of reagent in each instance:

$H_2N\text{—}G$ with $G = -NHC_6H_5$ or $-OH$ or $-NHCONH_2$

$F\text{-}\bigcirc\text{-}\underset{H}{C}=O + H_2NOH \longrightarrow F\text{-}\bigcirc\text{-}\underset{H}{C}=NOH + H_2O$

$F\text{-}\bigcirc\text{-}\underset{H}{C}=O + H_2NNHCONH_2 \longrightarrow F\text{-}\bigcirc\text{-}\underset{H}{C}=NNHCONH_2 + H_2O$

10.

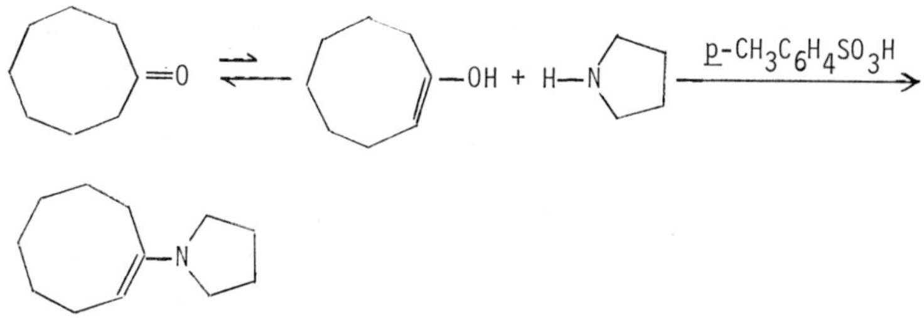

[cyclooctanone] =O ⇌ [cyclooctenol]—OH + H—N[pyrrolidine] $\xrightarrow{\text{p-CH}_3\text{C}_6\text{H}_4\text{SO}_3\text{H}}$

[cyclooctenyl-pyrrolidine]

an enamine

The reaction may be viewed as proceeding through the enol form. Enamines are important in modern synthetic organic chemistry.

11. 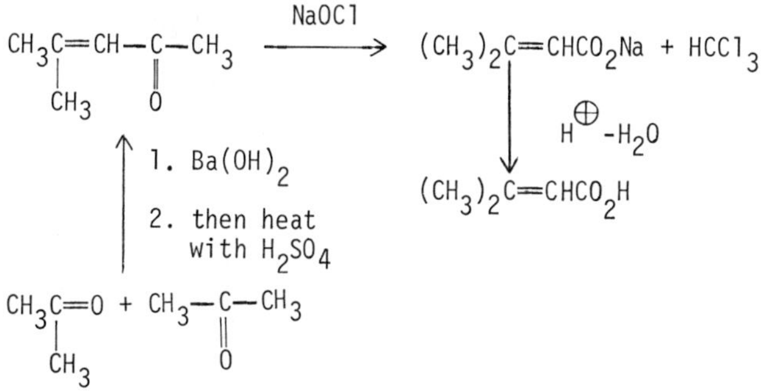
$$CH_3\underset{\underset{CH_3}{|}}{C}=CH-\underset{\underset{O}{\|}}{C}-CH_3 \xrightarrow{\text{NaOCl}} (CH_3)_2C=CHCO_2Na + HCCl_3$$

$$\downarrow H^{\oplus} -H_2O$$

$$(CH_3)_2C=CHCO_2H$$

↑ 1. Ba(OH)$_2$
  2. then heat with H$_2$SO$_4$

$$CH_3\underset{\underset{CH_3}{|}}{C}=O + CH_3-\underset{\underset{O}{\|}}{C}-CH_3$$

12.
$$H_2C=O \xrightarrow{\text{polymerize}} {+\!CH_2-O+\!}_n$$

$$H_2C=CH_2 \xrightarrow{\text{polymerize}} {+\!CH_2-CH_2+\!}_n$$

These are parallel types of polymerization reactions, but polyethylene is thermally more stable than polyformaldehyde.

13. 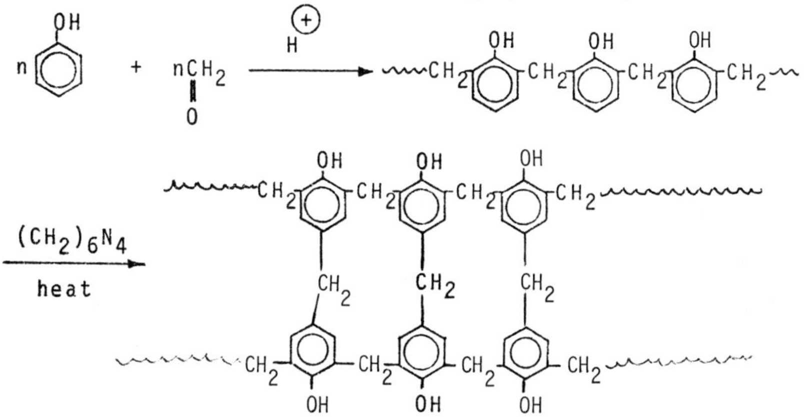
n[phenol]—OH + nCH$_2$=O $\xrightarrow{H^{(+)}}$ ∼∼CH$_2$—[phenol-OH]—CH$_2$—[phenol-OH]—CH$_2$—[phenol-OH]—CH$_2$∼∼

$\xrightarrow[\text{heat}]{(CH_2)_6N_4}$

∼∼CH$_2$—[phenol-OH]—CH$_2$—[phenol-OH]—CH$_2$—[phenol-OH]—CH$_2$∼∼
                  | | |
                  CH$_2$ CH$_2$ CH$_2$
                  | | |
∼∼CH$_2$—[phenol-OH]—CH$_2$—[phenol-OH]—CH$_2$—[phenol-OH]—CH$_2$∼∼

Note that this is another case of a carbonyl group reacting with an aromatic nucleus.

14.

~~~CH$_2$—CH—CH$_2$—CH—CH$_2$—CH—CH$_2$—CH~~~ $\xrightarrow{OH^{\ominus} \; -H_2O}$
         | | | |
         OAc   OAc   OAc   OAc

~~~CH$_2$—CH—CH$_2$—CH—CH$_2$—CH—CH$_2$—CH~~~ $\xrightarrow[H^{\oplus}]{CH_3CH_2CH_2C(H)=O}$
         | | | |
         OH   OH   OH   OH

[—CH$_2$—CH(O)—CH$_2$—CH(O)—] → [—CH$_2$—CH(O—CH(C$_3$H$_7$)—O)—CH$_2$—CH(O—CH(C$_3$H$_7$)—O)—]$_n$
       | |
       CHOH   CHOH
       | |
       C$_3$H$_7$ C$_3$H$_7$

butacite

15. (CH$_3$)$_2$C=CHCH$_2$CH$_2$\ /H   (CH$_3$)$_2$C=CHCH$_2$CH$_2$\ /CHO
                           C=C   C=C
                       H$_3$C/ \CHO   H$_3$C/ \H

         geranial   neral

16.

(1) [steroid structure with CH$_2$OH, C=O, OH groups and ring O]

(2) [steroid structure with CH$_2$OH, C=O, OH, CH$_3$, HO, F substituents]

17. (1) C$_6$H$_5$—C(=O)—C$_6$H$_5$

(2) di-sec-butyl ketone, 3,5-dimethyl-4-heptanone

(3) [cycloheptenone structure] =O

(4) 6,6-dimethyl-4-phenyl-2,5-heptanedione   (5) 7-octenal

(6) CH$_3$—⟨cyclohexyl⟩—CHO   (7) 2-methoxy-3-methylpentanal
                                       or
                                     3-methylvaleraldehyde

(8) CH$_3$—C(=O)—C$_6$H$_5$

18.

(1) 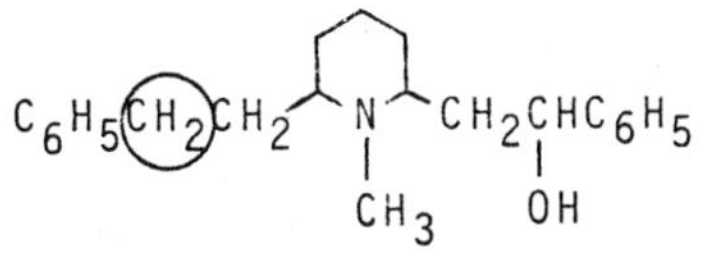 C₆H₅(CH₂)CH₂—N—CH₂CHC₆H₅ with CH₃ on N and OH on CH

(2) 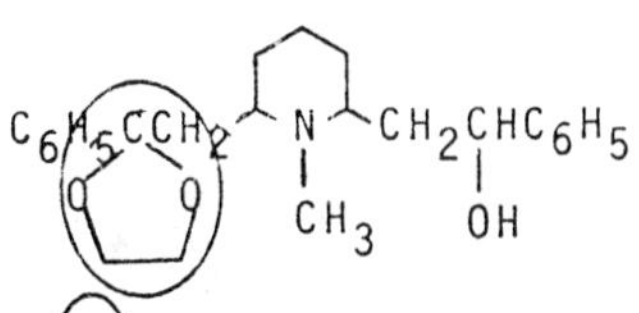 C₆H₅CCH₂—N—CH₂CHC₆H₅ (with cyclic ketal -O-O- circled) CH₃ on N, OH

(3) 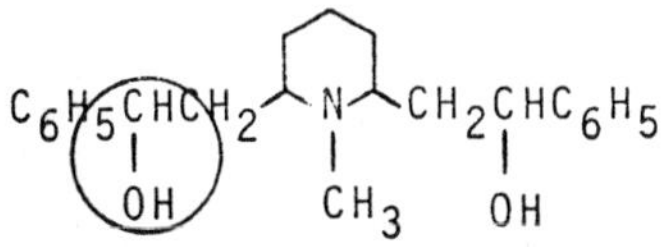 C₆H₅CHCH₂—N—CH₂CHC₆H₅ with (OH) circled, CH₃ on N, OH

(4) 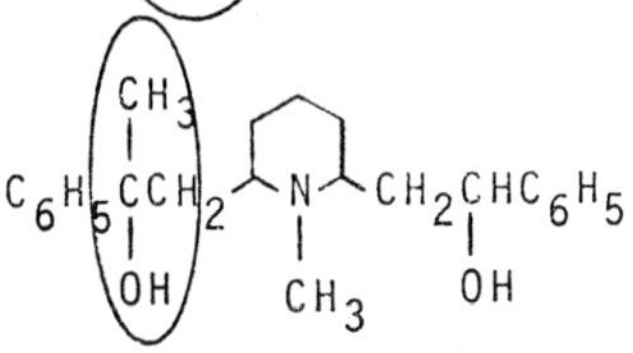 C₆H₅CCH₂—N—CH₂CHC₆H₅ with CH₃ and OH circled on left carbon, CH₃ on N, OH

(5) 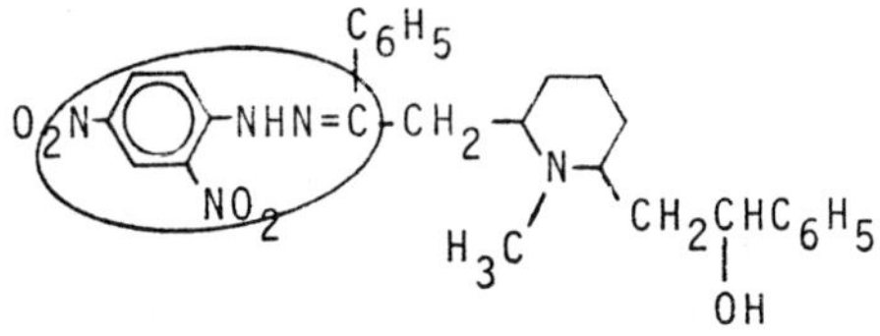 O₂N—C₆H₃(NO₂)—NHN=C(C₆H₅)—CH₂—[piperidine N with H₃C]—CH₂CHC₆H₅ with OH

These problems illustrate the reaction of a functional group in a more complex molecule. The modification which has occurred in the functional group is circled in each case. Recognize that a problem might occur with the -OH group in (1), (2), and (4).

19.

(1) 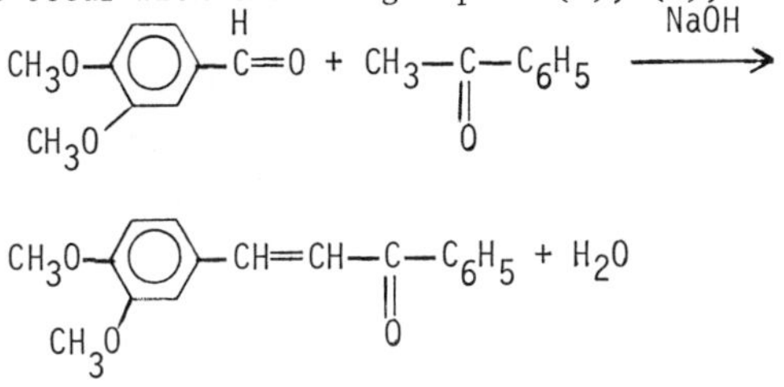 CH₃O—C₆H₃(OCH₃)—CH=O + CH₃—C(=O)—C₆H₅ $\xrightarrow{\text{NaOH}}$

CH₃O—C₆H₃(OCH₃)—CH=CH—C(=O)—C₆H₅ + H₂O

(2) 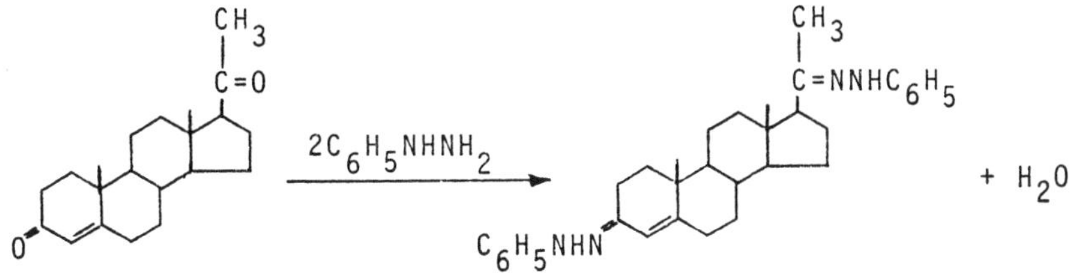 [steroid with C(CH₃)=O at top and =O at bottom] $\xrightarrow{2 C_6H_5NHNH_2}$ [steroid with C(CH₃)=NNHC₆H₅ and C₆H₅NHN=] + H₂O

(3) 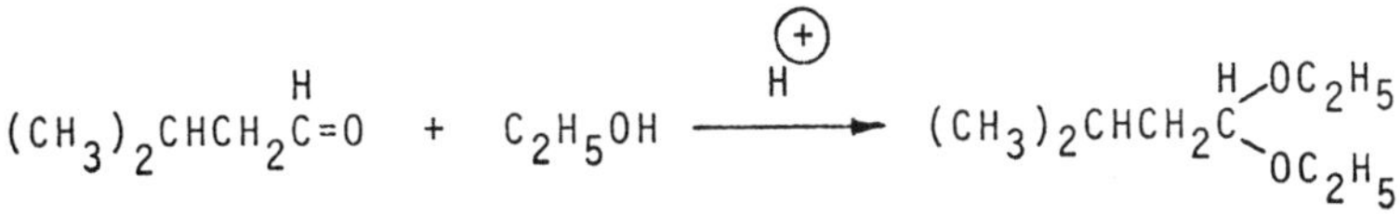 (CH₃)₂CHCH₂—CH=O + C₂H₅OH $\xrightarrow{H^+}$ (CH₃)₂CHCH₂—CH(OC₂H₅)₂

(4) 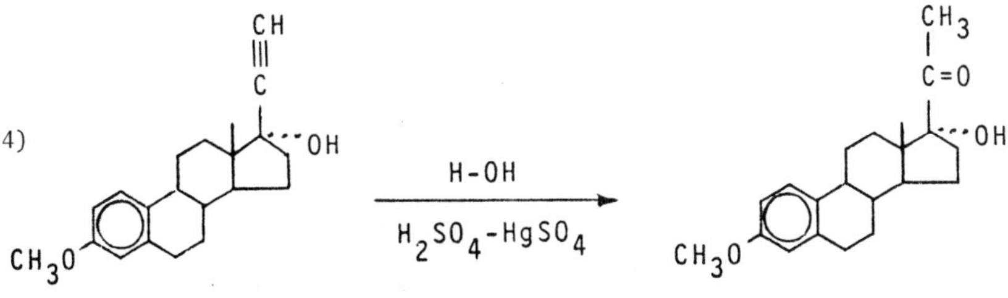 [steroid with C≡CH and OH, CH₃O on ring] $\xrightarrow[H_2SO_4-HgSO_4]{H-OH}$ [steroid with C(CH₃)=O and OH, CH₃O on ring]

113

(5) $CH_3(CH_2)_5CH_2OH \xrightarrow[300°]{Cu} CH_3(CH_2)_5CHO$

(6) $CH_3(CH_2)_3\underset{\underset{OH}{|}}{C}HCH_3 \xrightarrow{NaOI} CH_3(CH_2)_3CO_2Na + HCI_3\downarrow$

(7) 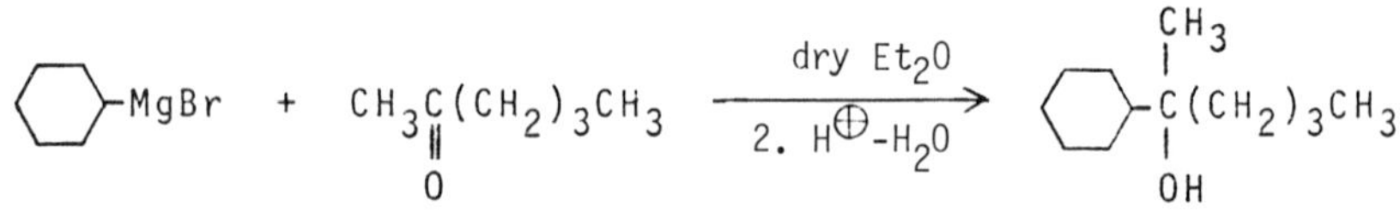 ⟨cyclohexyl⟩—MgBr + $CH_3\underset{\underset{O}{\|}}{C}(CH_2)_3CH_3 \xrightarrow[2.\ H^{\oplus}-H_2O]{dry\ Et_2O}$ ⟨cyclohexyl⟩—$\underset{\underset{OH}{|}}{\overset{\overset{CH_3}{|}}{C}}(CH_2)_3CH_3$

(8) $[CH_3(CH_2)_3\underset{\underset{O}{\|}}{C}-O-]_2 Fe \xrightarrow{heat} CH_3(CH_2)_3\underset{\underset{O}{\|}}{C}(CH_2)_3CH_3 + FeO + CO_2$

(9) 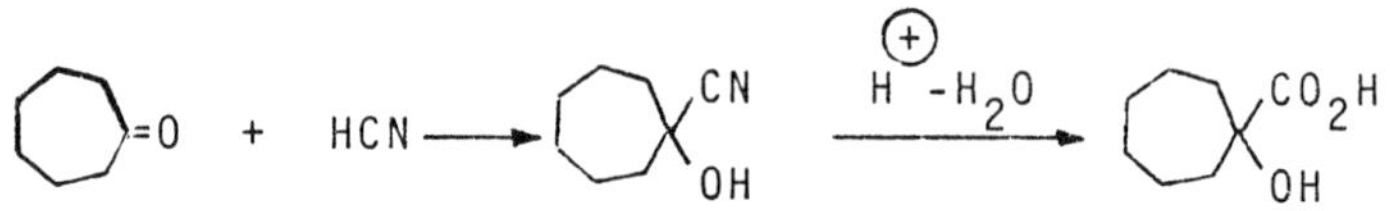 ⟨cycloheptanone⟩=O + HCN ⟶ ⟨cycloheptane⟩(CN)(OH) $\xrightarrow{H^{\oplus}-H_2O}$ ⟨cycloheptane⟩(CO_2H)(OH)

(10) 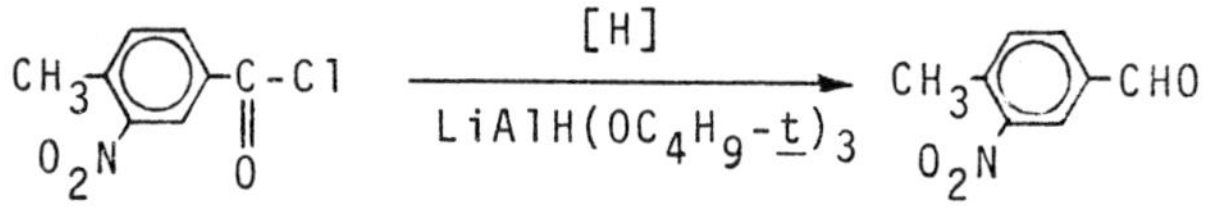 CH_3—⟨benzene, O_2N⟩—$\underset{\underset{O}{\|}}{C}-Cl \xrightarrow[LiAlH(OC_4H_9\text{-}\underline{t})_3]{[H]} CH_3$—⟨benzene, O_2N⟩—CHO

(11) 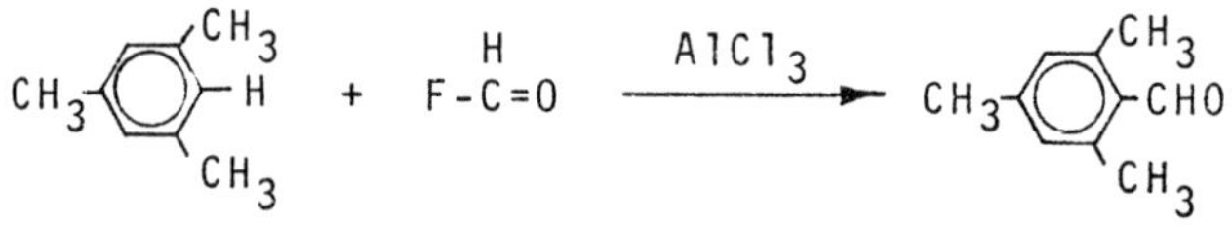 CH_3—⟨benzene, CH_3, CH_3⟩—H + F—C=O $\xrightarrow{AlCl_3}$ CH_3—⟨benzene, CH_3, CH_3⟩—CHO
               H

(12) $2(CH_3)_3C-CHO \xrightarrow{NaOH} (CH_3)_3CCH_2OH + (CH_3)_3C-CO_2Na$

20. 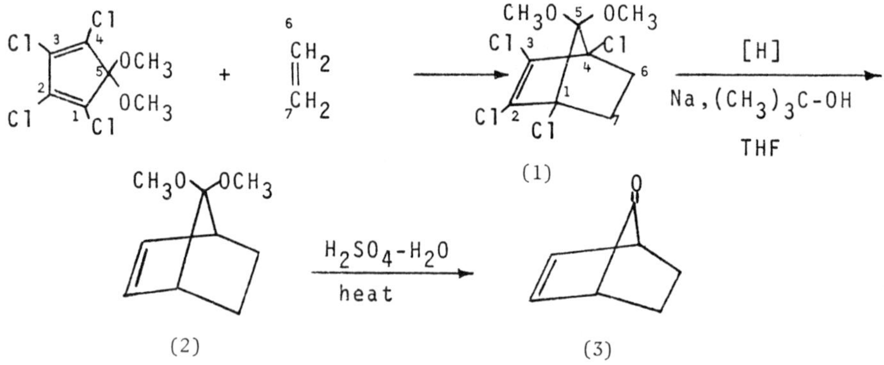 ⟨cyclopentadiene with Cl at 1,2,3,4 and OCH_3, OCH_3 at 5⟩ + $\underset{CH_2}{\overset{CH_2}{\|}}$ ⟶ ⟨bicyclic (1) with CH_3O, OCH_3 at 5; Cl at 1,2,3,4; C6, C7⟩ $\xrightarrow[Na,(CH_3)_3C-OH]{[H]}$
                                                                                                                                                                        THF

⟨bicyclic (2) with CH_3O, OCH_3⟩ $\xrightarrow[heat]{H_2SO_4-H_2O}$ ⟨bicyclic (3) with C=O⟩

21.
(1) $CH_3CH_2 \xrightarrow[\text{psi}]{800°} H_2C=CH_2 \xrightarrow[\text{2. }H_2O]{\text{1. }H_2SO_4} CH_3CH_2OH$

$\xrightarrow[275°]{Cu} CH_3\overset{H}{C}=O \xrightarrow{3Cl_2} Cl_3C-\overset{H}{C}=O$

$CH_3(CH_2)_4CH_3 \xrightarrow[\substack{500 \text{ psi} \\ 475°}]{Pt} \bigcirc \xrightarrow{Cl_2, Fe} \bigcirc-Cl \xrightarrow[H_2SO_4]{Cl_3C-CHO}$

$Cl_3C-CH-(\bigcirc-Cl)_2 \quad DDT$

(2) petroleum cracking $\xrightarrow[500°]{A_2O_3-SiO_2} CH_3CH=CH_2$

$\xrightarrow[H_2O]{H_2SO_4} CH_3\underset{OH}{CHCH_3} \xrightarrow[300°]{CuO} CH_3-\underset{\overset{\|}{O}}{C}-CH_3 \xrightarrow{HCN}$

$CH_3-\underset{\underset{OH}{|}}{\overset{\overset{CH_3}{|}}{C}}-CN \xrightarrow[CH_3OH]{H^{\oplus}} CH_2=\underset{CO_2CH_3}{\overset{\overset{CH_3}{|}}{C}} \xrightarrow[\text{heat}]{(RO\rightarrow)_2} \left[CH_2-\underset{CO_2CH_3}{\overset{\overset{CH_3}{|}}{C}} \right]_n$

$\xrightarrow[\substack{4000-5000 \text{ psi}}]{\substack{350° \\ ZnO-Cr_2O_3}}$

$CH_4 \xrightarrow{O_2} CO + 2H_2$

Note how any shortage of petroleum would have a major impact on the production of organic chemicals and consumer products made therefrom.

22.
(1) $C_2H_5-\bigcirc \xrightarrow[AlCl_3]{HCOF} C_2H_5-\bigcirc-CHO \xrightarrow[H^{\oplus}]{C_2H_5OH} C_2H_5-\bigcirc-CH(OC_2H_5)_2$

(2) $(CH_3)_2CHCH_2CH_2OH \xrightarrow[300°]{Cu} (CH_3)_2CHCH_2\overset{H}{C}=O, \quad C_6H_5CH_2OH \xrightarrow{HBr}$

$C_6H_5CH_2Br \xrightarrow[Mg]{\text{dry }Et_2O} C_6H_5CH_2MgBr \xrightarrow[\text{2. }H^{\oplus}-H_2O]{(CH_3)_2CHCH_2C=O}$

$(CH_3)_2CHCH_2\underset{OH}{\overset{|}{C}}HCH_2C_6H_5 \xrightarrow[300°]{Cu} (CH_3)_2CHCH_2\underset{\overset{\|}{O}}{C}CH_2C_6H_5$

(3) $(CH_3)_2C=O + CH_3CCH_3 \xrightarrow{\overset{\oplus}{OH}} (CH_3)_2C=CHCCH_3 \xrightarrow[Ni]{H_2}$
$\underset{O}{\|}\underset{O}{\|}$

$(CH_3)_2CHCH_2CHCH_3 \xrightarrow[300°]{Cu} (CH_3)_2CHCH_2CCH_3$
$|\|$
OHO

(4) $\triangledown\!O \xrightarrow[2.\ H^{\oplus}-H_2O]{C_6H_5CH_2MgBr} C_6H_5CH_2CH_2CH_2OH \xrightarrow[300°]{Cu} C_6H_5CH_2CH_2\overset{H}{C}=O$

$\xrightarrow{HCN} C_6H_5CH_2CH_2\underset{OH}{\overset{|}{CH}}\!-\!CN \xrightarrow{1.\ H^{\oplus}-H_2O} C_6H_5CH_2CH_2\underset{OH}{\overset{|}{CH}}CO_2H$

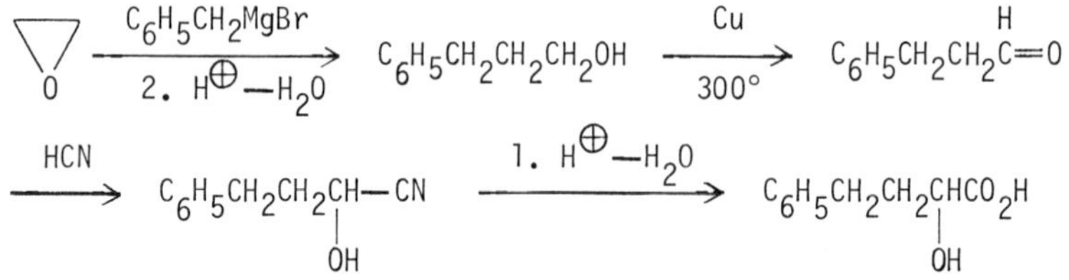

(5) ⌬-OH $\xrightarrow[NaCr_2O_7^-,\ H_2SO_4]{[O]}$ ⌬=O

(6) $C_6H_6 \xrightarrow[FeBr_3]{Br_2} C_6H_5Br \xrightarrow[Mg]{dry\ Et_2O} C_6H_5MgBr$

⌬=O $\xrightarrow{2.\ H^{\oplus}-H_2O}$ ⌬—⌬

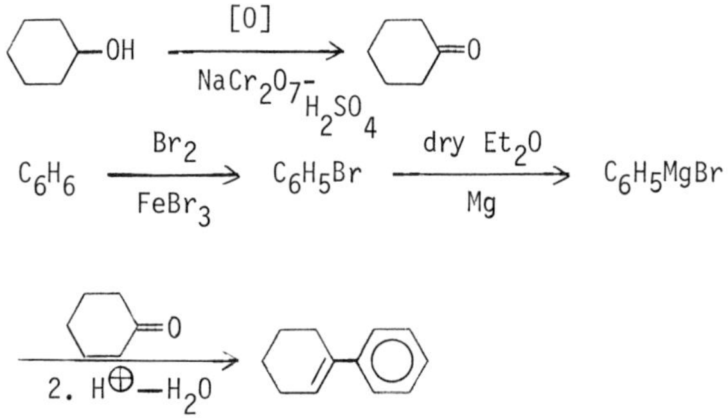

23. (1) This reaction should be compared with that of a Grignard reagent and a ketone:

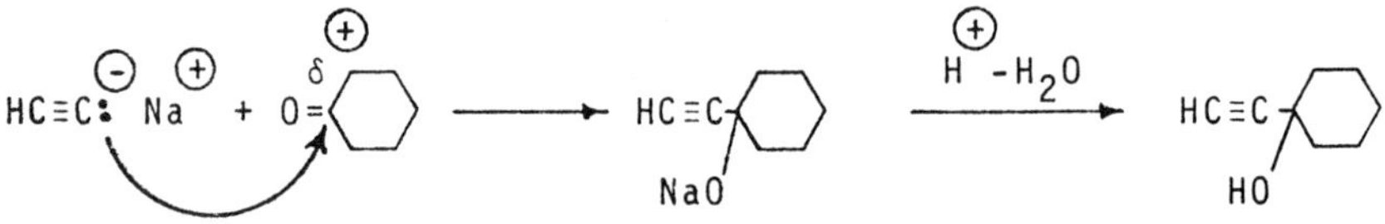

$HC\equiv C:^{\ominus} Na^{\oplus} + O=$⌬ $\longrightarrow HC\equiv C-$⌬$-NaO \xrightarrow{H^{+}-H_2O} HC\equiv C-$⌬$-HO$

(2) This reaction first involves an aldol condensation between the acetaldehyde and formaldehyde:

$CH_3-CHO \xrightarrow{:OH^{\ominus}} :CH_2C=O \xrightarrow{H_2C=O} :OCH_2CH_2C=O \xrightarrow{H_2O}$
HH
$+H_2O$

$HOCH_2CH_2CHO + OH^{\ominus} \xrightarrow[H_2C=O,\ OH^{\ominus}]{2\ more\ times} HOCH_2\underset{CH_2OH}{\overset{CH_2OH}{\overset{|}{C}}}\!-\!CHO \xrightarrow[OH^{\ominus}]{H_2C=O} HOCH_2\underset{CH_2OH}{\overset{CH_2OH}{\overset{|}{C}}}\!-\!CH_2OH + HCO_2Na$

The final step involves a crossed Cannizzaro reaction with formaldehyde going to the acid salt. The product is known as pentaerythritol [2,2-di-(hydroxymethyl)-1,3-propanediol] and is used in the manufacture of certain resins and the tetranitrate, a high explosive.

24. (1) Only acetaldehyde (a) reacts with sodium hypoiodite to give a milky precipitate of iodoform:

$$CH_3CHO + NaOI \longrightarrow I_3C-CHO \xrightarrow{OH^\ominus} CHI_3\downarrow + HCO_2Na$$

(2) Only the ketone (b) reacts with 2,4-dinitrophenylhydrazine to give a yellow-orange precipitate:

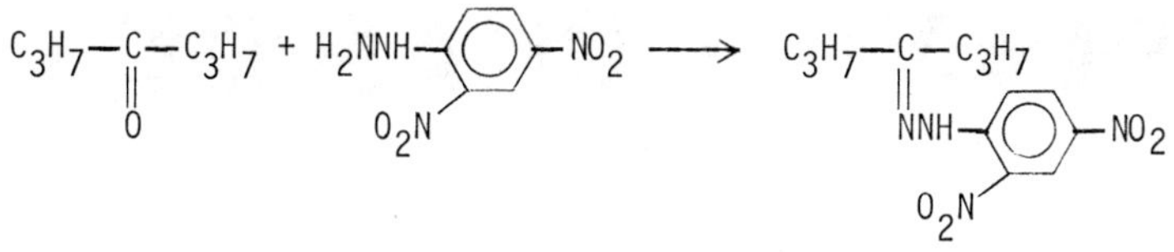

$$C_3H_7-\underset{O}{\overset{\|}{C}}-C_3H_7 + H_2NNH-\underset{O_2N}{\bigcirc}-NO_2 \longrightarrow C_3H_7-\underset{NNH-\underset{O_2N}{\bigcirc}-NO_2}{\overset{\|}{C}}-C_3H_7$$

(3) Only the aldehyde (b) will react with Tollen's reagent to give a silver mirror:

$$C_4H_9-CHO + [Ag_2O] \xrightarrow{Ag(NH_3)_2^\oplus} C_4H_9CO_2NH_4 + 2Ag\downarrow$$

(4) Only the methyl ketone (a) will react with sodium hypoiodite to give iodoform:

$$CH_3(CH_2)_3\underset{O}{\overset{\|}{C}}CH_3 \xrightarrow{NaOI} CH_3(CH_2)_3CO_2Na + CHI_3\downarrow$$

chapter 13

CARBOHYDRATES: SUGARS AND POLYSACCHARIDES

1. $HOCH_2\!-\!(CH)_3\!-\!C$ with a ring: $HC\!=\!N\!-\!N\!-\!C_6H_5$, $N\cdots H$, NHC_6H_5; with $-OH$ on the CH chain.

2. $HOCH_2CH_2OH \xrightarrow{Ac_2O\,(xs)} AcOCH_2CH_2OAc \qquad Ac = CH_3\overset{\underset{\parallel}{O}}{C}-$

3. C-2: H—, —CHO, —OH, $HOCH_2CH\!-\!CH\!-\!CH\!-$; C-3: H—, —CHCHO,
    $\qquad\qquad\qquad\qquad\qquad\quad\;\;\;\underset{OH}{|}\;\underset{OH}{|}\;\underset{OH}{|}\qquad\qquad\qquad\qquad\underset{OH}{|}$

    —OH, $HOCH_2CH\!-\!CH-$; C-4: H—, $-CH\!-\!CHCHO$, —OH, $HOCH_2CH-$;
    $\quad\;\;\underset{OH}{|}\;\underset{OH}{|}\qquad\qquad\qquad\quad\;\;\underset{OH}{|}\;\underset{OH}{|}\qquad\qquad\qquad\qquad\underset{OH}{|}$

    C-5: H—, $-CH\!-\!CH\!-\!CHCHO$, —OH, $HOCH_2-$
    $\qquad\qquad\;\underset{OH}{|}\;\underset{OH}{|}\;\underset{OH}{|}$

4.
    ```
         CHO                    CH=NNHC6H5              CH2OH
         |                      |                       |
      H—C—OH                    C==NNHC6H5              C=O
         |          C6H5NHNH2   |                       |
      HO—C—H       ─────────→   HO—C—H       ←─────     HO—C—H
         |                      |                       |
      H—C—OH                    H—C—OH                  H—C—OH
         |                      |                       |
      H—C—OH                    H—C—OH                  H—C—OH
         |                      |                       |
         CH2OH                  CH2OH                   CH2OH
    ```
    D(+)-glucose   D(−)-fructose

    (See top of page 119.)

```
        CHO
   HO—C—H
   HO—C—H
    H—C—OH
    H—C—OH
       CH₂OH
```
D(+)-mannose

All three of these sugars form an identical osazone. That is, once the configuration of glucose was known, the configurations of mannose and fructose were also known. Since the common osazone showed that the two aldohexoses, glucose and mannose, were identical in C_3-C_5, mannose could only be C_2 epimer of glucose.

5.
```
   CH₂OH                              CH=NNHC₆H₅
    |          C₆H₅NHNH₂(xs)           |
    C=O       ─────────────►           C=NNHC₆H₅
    |                                  |
   CH₂OH                              CH₂OH
```

6. (1)
```
   CH₂OH          (2)  CH₂OAc       (3)  CO₂H       (4)  CO₂H
    |                   |                 |                |
   HCOH                HCOAc             HOCH             HCOH
    |                   |                 |                |
   HOCH                AcOCH             HOCH             HOCH
    |                   |                 |                |
   HOCH                AcOCH             HCOH             HCOH
    |                   |                 |                |
   HCOH                HCOAc             HCOH             HCOH
    |                   |                 |                |
   CH₂OH               CH₂OAc            CO₂H             CH₂OH
```

7.

β-D-mannose

α-D-galactose

8.
$$CH_3-\underset{H}{C}=O \xrightarrow[H^{\oplus}]{HOCH_3} \left[CH_3-\underset{\underset{1}{OH}}{\overset{H}{\underset{|}{C}}}\underset{2}{-}O\underset{3}{CH_3} \right] \xrightarrow[H^{\oplus}]{HOCH_3} CH_3-\underset{OCH_3}{\overset{H}{\underset{|}{C}}}-OCH_3$$

unstable hemiacetal   an acetal

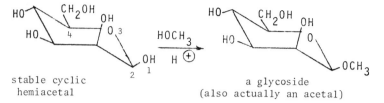

| stable cyclic hemiacetal | | a glycoside (also actually an acetal) |

Because of the stability and ease of formation of six-membered rings, it is logical that, although the open-chain hemiacetal is unstable, the corresponding six-membered cyclic hemiacetal is stable. The hemacetal group is numbered for both open-chain and cyclic structures.

9. 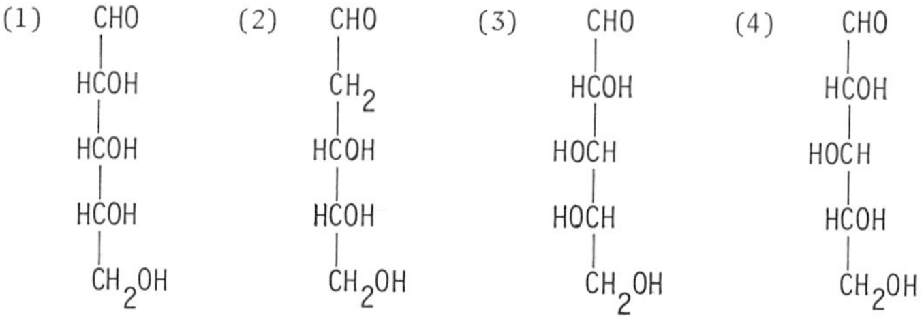 (1) CHO
    HCOH
    HCOH
    HCOH
    CH_2OH

    (2) CHO
    CH_2
    HCOH
    HCOH
    CH_2OH

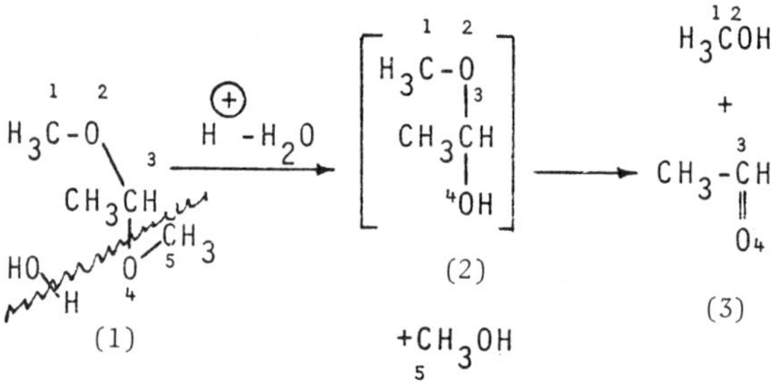

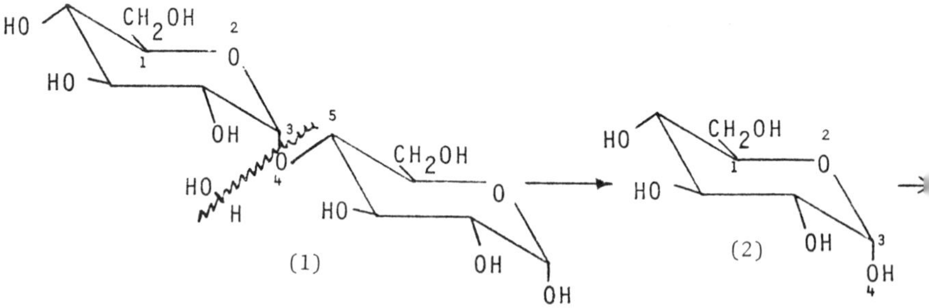

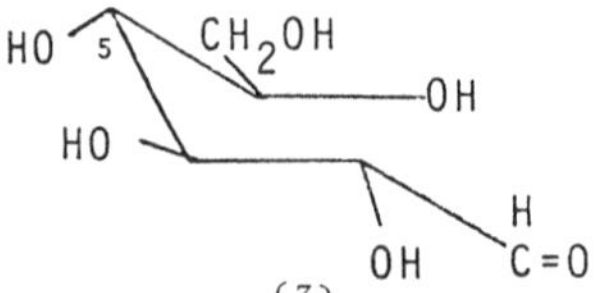

    (3) CHO
    HCOH
    HOCH
    HOCH
    CH_2OH

    (4) CHO
    HCOH
    HOCH
    HCOH
    CH_2OH

10.

It should be observed that these sequences are exactly parallel at each stage.

120

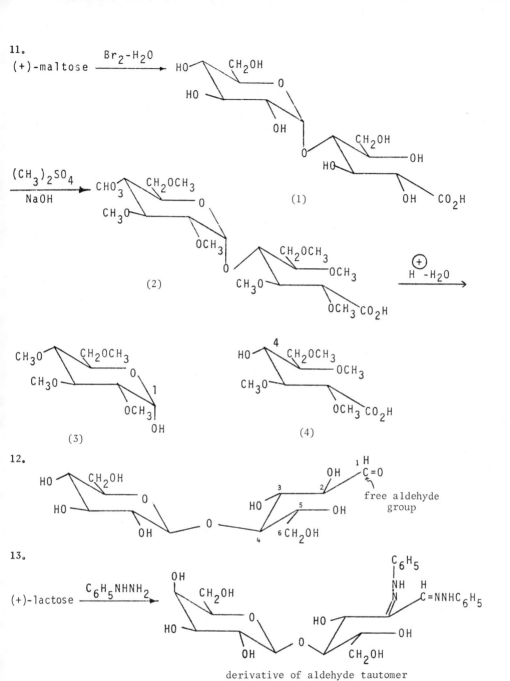

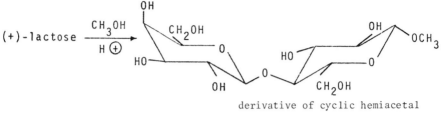

14.

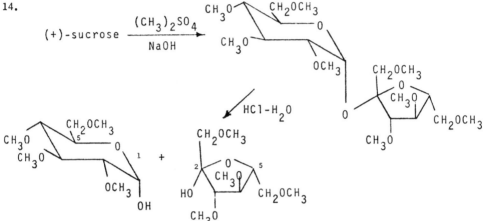

In these experiments, methyl groups are missing at C-1 and C-5 in the glucose portion, indicating a pyranose ring, and at C-2 and C-5 in the fructose portion, indicating a furanose ring.

15.

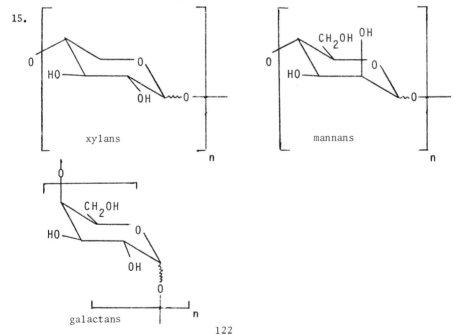

Observe that the notation used in these formulas is a noncommittal representation suggesting that the glycosidic linkage can be either α- or β- but not specifying which.

16.

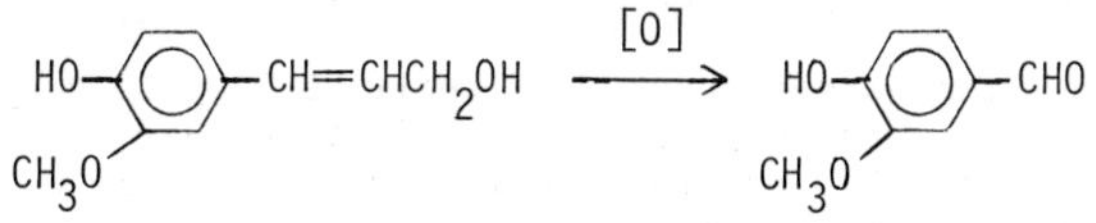

HO—⟨C₆H₃(OCH₃)⟩—CH=CHCH₂OH $\xrightarrow{[O]}$ HO—⟨C₆H₃(OCH₃)⟩—CHO

coniferyl alcohol → vanillin

17. 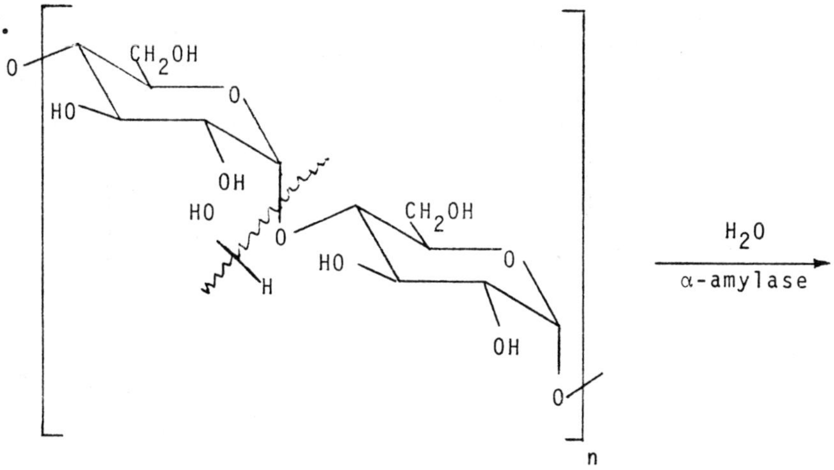 starch (amylose) $\xrightarrow[\alpha\text{-amylase}]{H_2O}$

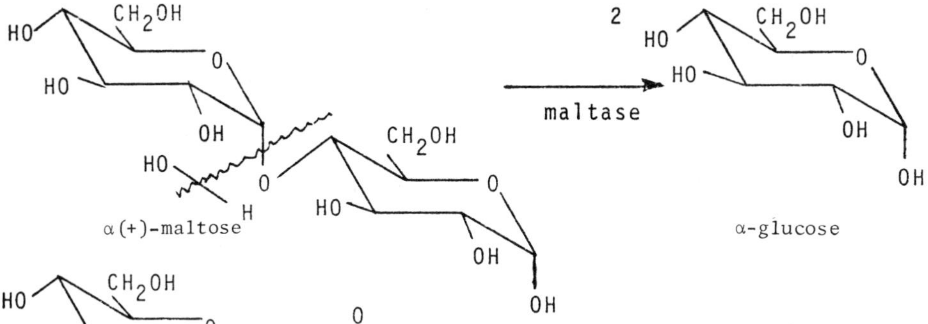

α(+)-maltose $\xrightarrow{\text{maltase}}$ 2 α-glucose

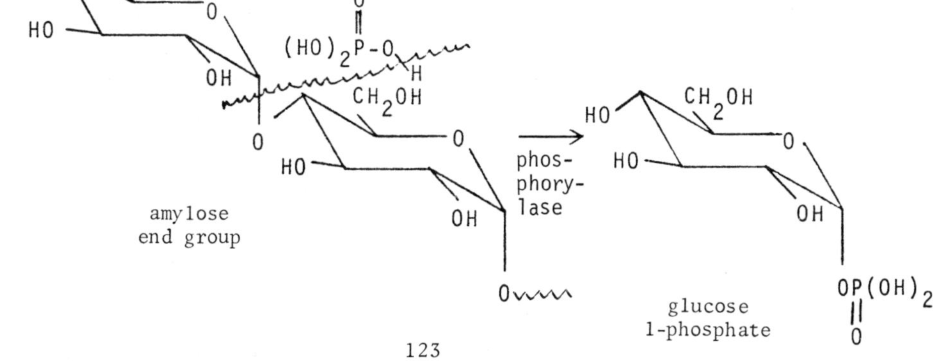

amylose end group $\xrightarrow{\text{phos-phory-lase}}$ glucose 1-phosphate

In phosphorolysis, terminal glucose units are split off as glucose-1-phosphate units. Thus, glucose is readily available on demand from its storage form, glycogen. The reverse reaction also occurs and represents the formation of glycogen from glucose.

18. The reducing action of sugars is caused by their ability to chain tautomerize to give an aldehyde group [as in (1)]. Further, a keto-hexose, such as fructose (2), can tautomerize to an aldohexose, glucose (1), and thus indirectly furnish an aldehyde group.

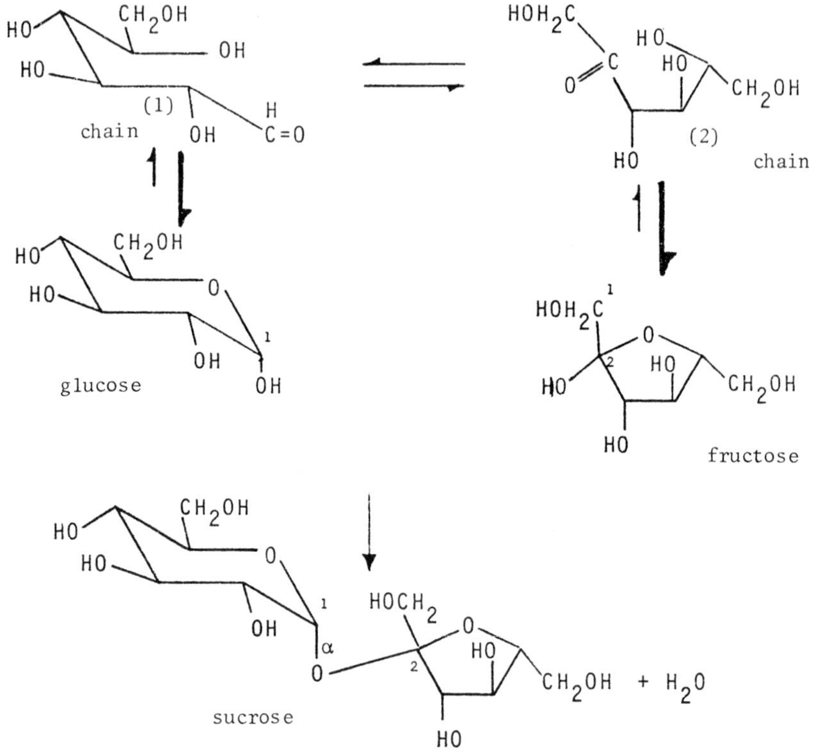

Sucrose can be viewed as having been formed from one equivalent of glucose and one of fructose by the elimination of a molecule of water through the hemiacetal hydroxyl groups on C-1 in glucose and on C-2 in fructose. This now leaves no hemiacetal hydroxyl through which a ring can open to give the chain tautomer containing the aldehyde group. Note that a hemiacetal hydroxyl group is the only hydroxyl group in a monosaccharide which is attached to a carbon that is attached in turn to the oxygen of the ring.

19. 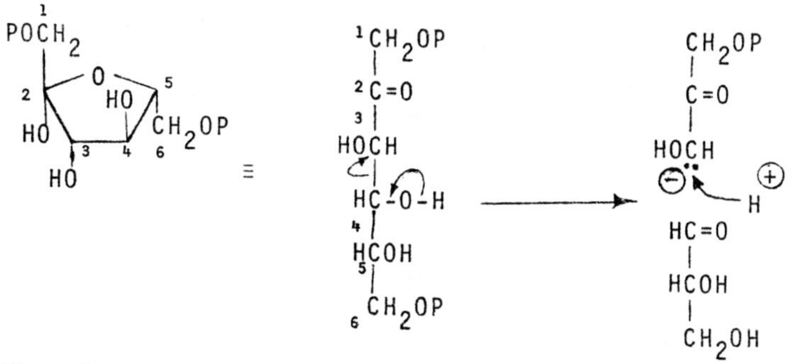 [Structure of phosphorylated sugar shown in ring form with positions 1-6 labeled: POCH₂ at 1, HO at 2, HO at 3, ring O between 4 and 5, CH₂OP at 6] ≡ [Open chain form: ¹CH₂OP, ²C=O, ³HOCH, ⁴HC–O–H, ⁵HCOH, ⁶CH₂OP] ⟶ [CH₂OP, C=O, HOCH⊖ ⋯ H⊕, HC=O, HCOH, CH₂OH]

The anion of the dihydroxyacetone phosphate immediately takes on the proton.

20. (1) [Chair form sugar with HO, OH groups and OH at anomeric position] ⇌ [Open chain form with HO, OH groups and terminal H–C=O]

$C_6H_5NHNH_2$ ⟶ [Osazone chain structure with HO, OH, C=NNHC₆H₅ and N–NH–C₆H₅]

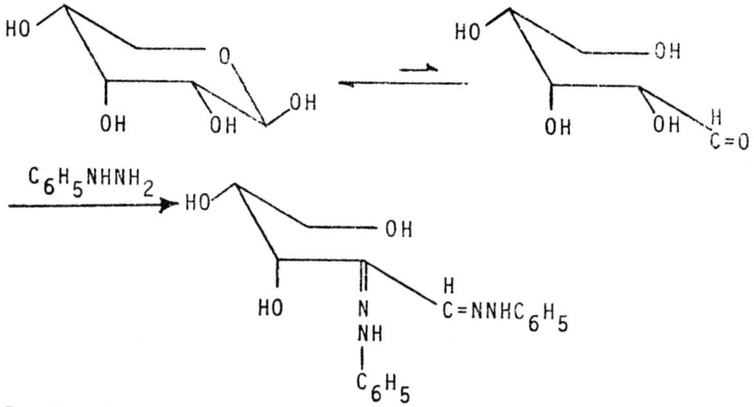

In all of these cases, the chain formula will be shown for clarity.

(2) [Disaccharide peracetate structure with OAc, CH₂OAc, AcO groups on two linked pyranose rings] $Ac = CH_3\underset{\underset{O}{\|}}{C}-$

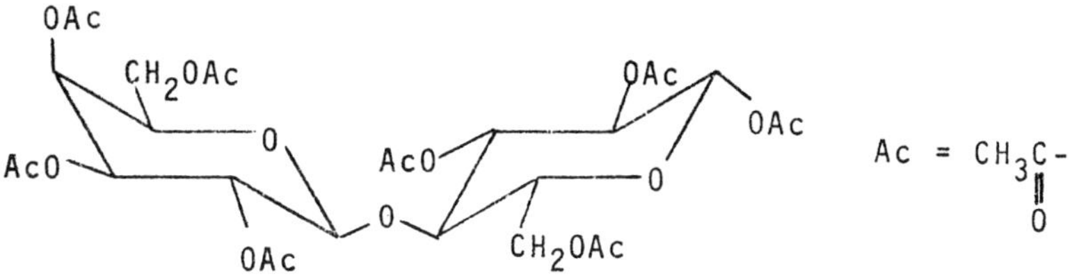

(3) [Pyranose ring with OH, HO, OH groups] ⇌ [Open chain form with OH groups and H–C=O] $\xrightarrow[HNO_3]{[O]}$ [COOH, HCOH, HOCH, HOCH, COOH]

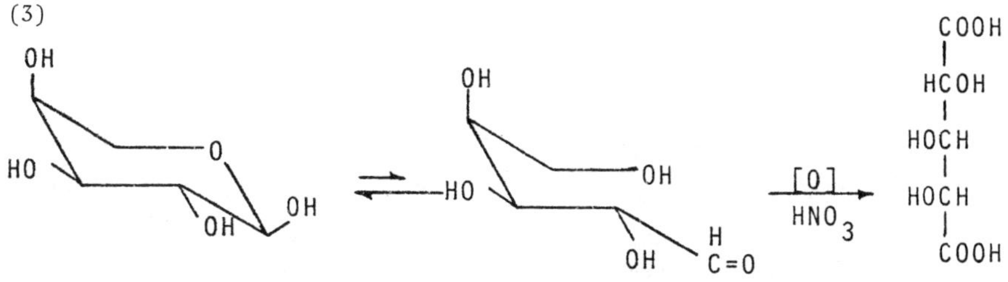

Note that the chain formula for the dicarboxylic acid on the right is no different in type than the chain formula of the osazone written in (1). The way in which the latter is written relates closely to the ring formula. We should gain experience in interrelating such formulas.

(4)
```
  CN              CN          (5)  CH₂OH
  |               |                |
 HCOH           HOCH              HOCH
  |               |                |
 HCOH           HCOH              HOCH
  |               |                |
 HOCH    +      HOCH              HCOH
  |               |                |
 HOCH           HOCH              HCOH
  |               |                |
 HCOH           HCOH              CH₂OH
  |               |
 CH₂OH          CH₂OH
```

(6)

$$\text{HOCH}_2\text{-ring with OH, O, HO} \rightleftharpoons \text{HOCH}_2\text{-open form with } \overset{H}{\underset{}{C=O}}, \text{OH, HO} \xrightarrow[Br_2-H_2O]{[O]} \text{HOCH}_2\text{-form with OH, HO, } CO_2H$$

21.

(sugar ring with HO, HO, CH₂OH, O, OH, OH) $+$ $HO-\overset{O}{\underset{OH}{\overset{\|}{P}}}-O-\overset{O}{\underset{OH}{\overset{\|}{P}}}-O-\overset{O}{\underset{OH}{\overset{\|}{P}}}-OCH_2$ (ribose-adenine nucleoside with HO, OH, NH₂, N-ring)

$\longrightarrow$ (sugar ring with HO, HO, CH₂OP(OH)₂, O, OH, OH) $+$ $HO-\overset{O}{\underset{OH}{\overset{\|}{P}}}-O-\overset{O}{\underset{OH}{\overset{\|}{P}}}-OA$

A = adenosine residue. This is essentially the reaction of an alcohol with an anhydride, as is ethanol with acetic anhydride.

$$CH_3CH_2OH \; + \; CH_3\overset{O}{\overset{\|}{C}}-O-\overset{O}{\overset{\|}{C}}CH_3 \longrightarrow CH_3CH_2O\overset{O}{\overset{\|}{C}}CH_3 \; + \; HO\overset{O}{\overset{\|}{C}}CH_3$$

22. These inorganic esters may be viewed as being derived in the same way through a hydroxyl group and the appropriate inorganic acid.

$$CH_3(CH_2)_{10}CH_2O-H + HO-SO_2OH \longrightarrow CH_3(CH_2)_{10}CH_2OSO_2OH + H_2O$$

  sulfuric acid   lauryl hydrogen sulfate,
                       a sulfate ester

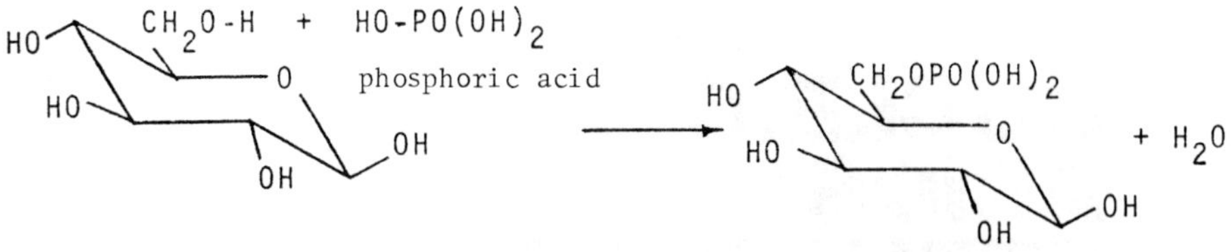

$$\text{HO-glucose} + \text{HO-PO(OH)}_2 \text{ (phosphoric acid)} \longrightarrow \text{glucose-6-phosphate, a phosphate ester} + H_2O$$

23.

$$\begin{array}{c} CHO \\ | \\ CH_2 \\ H \end{array} + \begin{array}{c} HC=O \\ | \\ CH_3 \end{array} \xrightarrow{OH^-} \begin{array}{c} CHO \\ | \\ CH_2 \\ | \\ HCOH \\ | \\ CH_3 \end{array}$$

$$\begin{array}{c} CH_2OPO(OH)_2 \\ | \\ C=O \\ | \\ HOCH \\ H \end{array} + \begin{array}{c} HC=O \\ | \\ HCOH \\ | \\ CH_2OPO(OH)_2 \end{array} \xrightleftharpoons{\text{aldolase}} \begin{array}{c} CH_2OPO(OH)_2 \\ | \\ C=O \\ | \\ HOCH \\ | \\ HCOH \\ | \\ HCOH \\ | \\ CH_2OPO(OH)_2 \end{array}$$

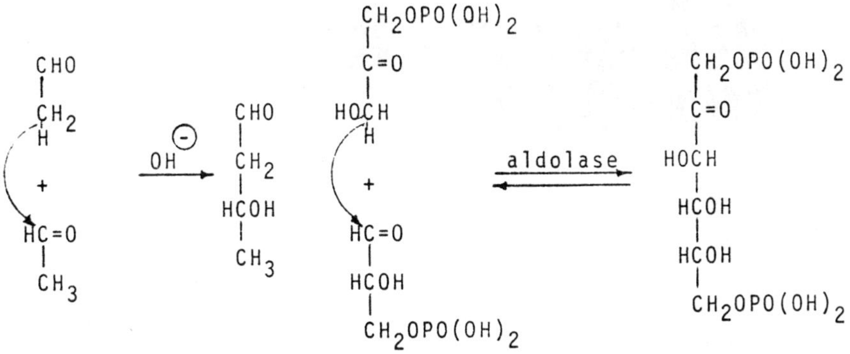

Both of these aldol condensations are identical in principle. Acetaldehyde is a simple model for the more complex structures in the second case. The α-hydrogen activated by the adjacent carbonyl group in a second molecule of acetaldehyde, whereas the α-carbon atom forms a bond with the carbon of the carbonyl group. The fructose-1,6-diphosphate formation involves the reaction of α-carbon and carbonyl groups catalyzed by the enzyme aldolase.

24.

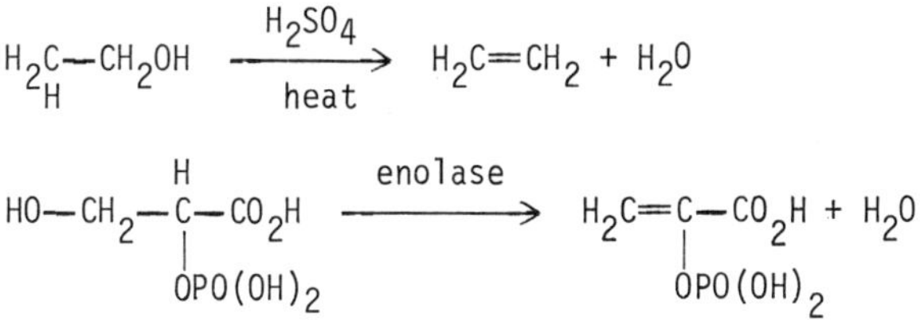

$$H_2\underset{H}{C}\text{—}CH_2OH \xrightarrow[\text{heat}]{H_2SO_4} H_2C\text{=}CH_2 + H_2O$$

$$HO\text{—}CH_2\text{—}\underset{OPO(OH)_2}{\overset{H}{C}}\text{—}CO_2H \xrightarrow{\text{enolase}} H_2C\text{=}\underset{OPO(OH)_2}{C}\text{—}CO_2H + H_2O$$

25.

pyran   furan

chapter 14

NUCLEIC ACIDS: STRUCTURE AND MOLECULAR BIOLOGY

1.

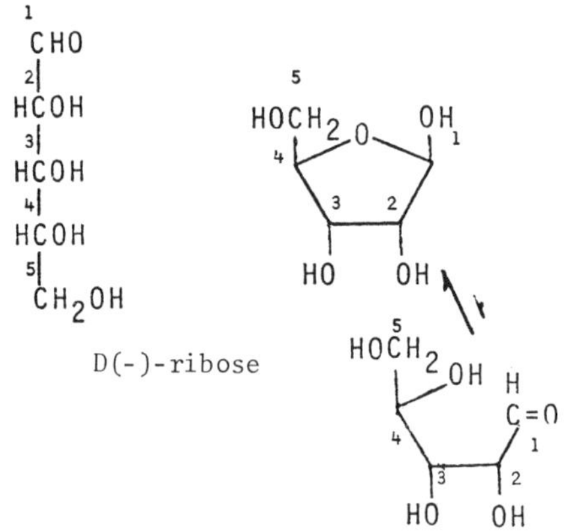

```
  1
  CHO
  2|
  HCOH
  3|
  HCOH
  4|
  HCOH
  5|
  CH2OH

  D(-)-ribose
```

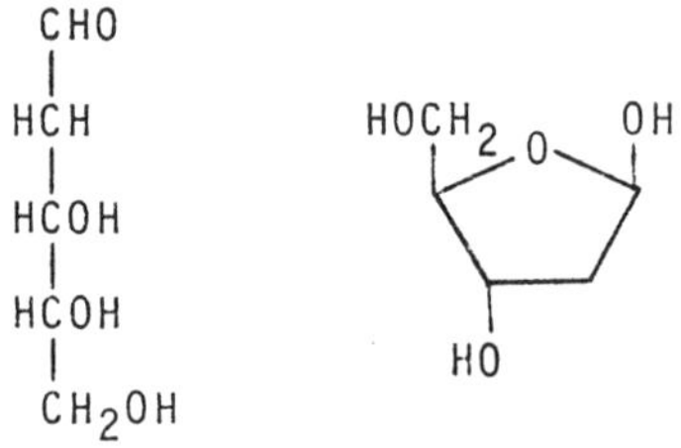

[ribose furanose ring with 5-HOCH$_2$, O, 1-OH, 3-HO, 2-OH]

```
  CHO
  |
  HCH
  |
  HCOH
  |
  HCOH
  |
  CH2OH
```

[2-deoxy ribose furanose ring with 5-HOCH$_2$, O, 1-OH, 3-HO]

2-deoxy-D-ribose

[open chain form with 5-HOCH$_2$, 4, 3-HO, 2-OH, 1 C=O, H]

The hydroxyl and aldehyde groups then interact to give a cyclic hemiacetal.

Chain structure is written here to conform with the five-membered ring.

2. Write and relate the formulas in 11.30 and 14.2.

3. In positions 2 or 4 of pyridine, the O—H bond will interact through the ring with the ring N to give the stable lactam tautomer. In position 3, this type of interaction is impossible. Unlike the hydroxyl group in positions 2 and 4, the amino group does not tautomerize to a stable tautomer. These tautomeric considerations are important in respect to the structure of nucleic acids.

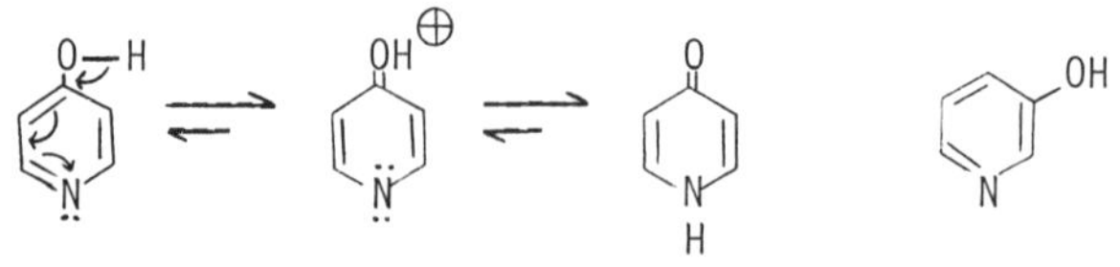

[Four pyridine tautomer structures shown: unstable (O—H form with arrows), intermediate (OH$^{\oplus}$ form), stable (lactam, C=O with N—H), and only structure possible (3-OH pyridine)]

    unstable   stable   only structure possible stable ⇌ unstable

4.

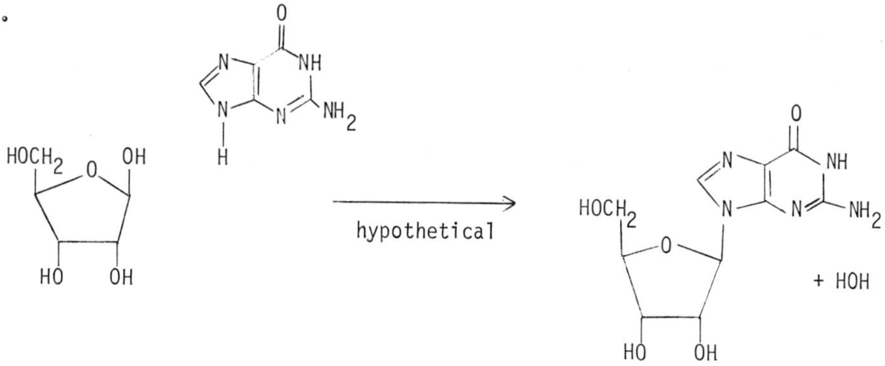

[ribose] + guanine →(hypothetical) guanosine + HOH

guanosine, a nucleoside or N-glycoside

Nucleosides cannot be prepared by the direct approach used for glycosides; rather, more involved special synthetic methods are required.

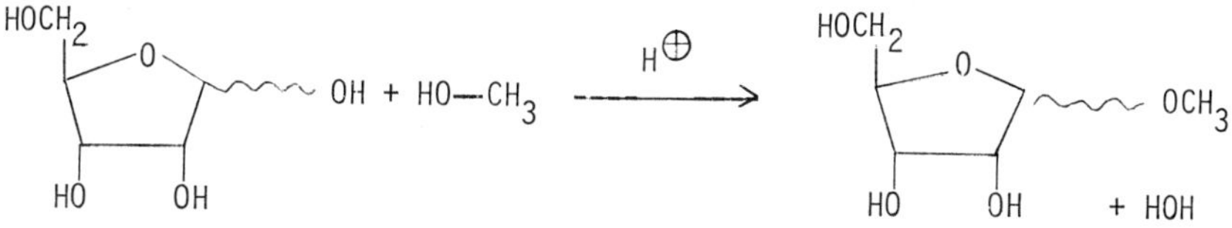

α- or β-D-ribose ~OH + HO—CH₃ →(H⊕) α- or β-D-riboside, a glycoside ~OCH₃ + HOH

5.

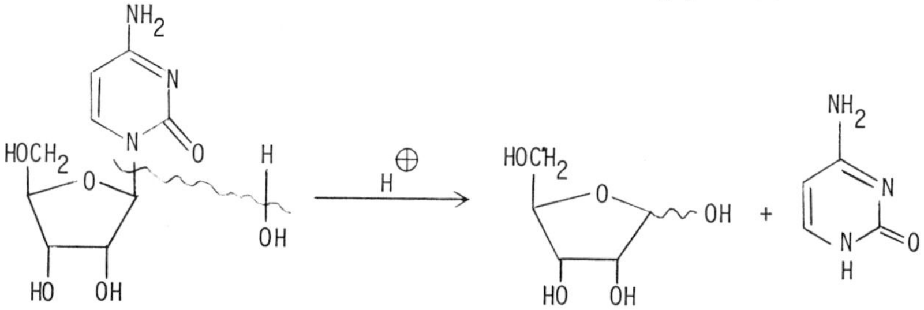

[cytidine] →(H⊕, H₂O) D-ribose + cytosine

[ribose-OC₂H₅] →(H⊕) D-ribose + ethanol

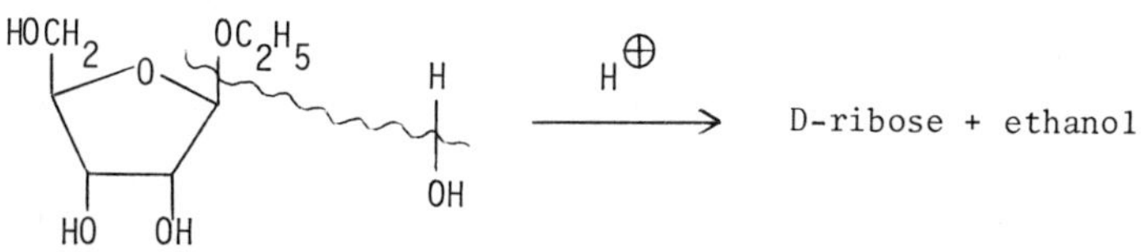

For maltose, see Pb. 13.10. The acidic hydrolysis of all three is identical in principle, the hydrolytic cleavage of an acetal-like structure.

6.

(1) $CH_3(CH_2)_3O-H + HO-\underset{\underset{O}{\|}}{\overset{\overset{O}{\|}}{S}}-OH \dashrightarrow CH_3(CH_2)_3O-\underset{\underset{O}{\|}}{\overset{\overset{O}{\|}}{S}}-OH + HOH$

(2) $CH_3(CH_2)_3O-H + HO-NO_2 \dashrightarrow CH_3(CH_2)_3O-NO_2 + HOH$

(3) $CH_3(CH_2)_3O-H + HO-\underset{\underset{OH}{|}}{\overset{\overset{O}{\|}}{P}}-OH \dashrightarrow CH_3(CH_2)_3O-PO(OH)_2 + HOH$

(4) $CH_3(CH_2)_3O-H + HO-\underset{\underset{O}{\|}}{C}CH_3 \dashrightarrow CH_3(CH_2)_3O-\underset{\underset{O}{\|}}{C}CH_3 + HOH$

All four reactions are identical in principle, examples of ester formation. The first three are esters of inorganic acids, and the the fourth represents the large class of organic esters. Examples (3) and (1) illustrate phosphate and sulfate esters, both common in biological systems.

7. (1) (2) 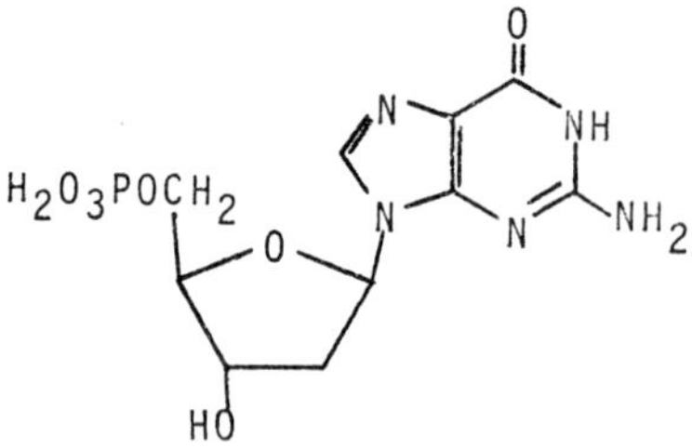 (3) 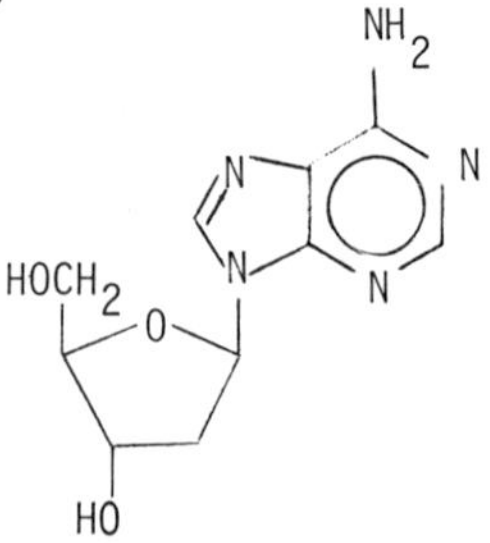 (4)

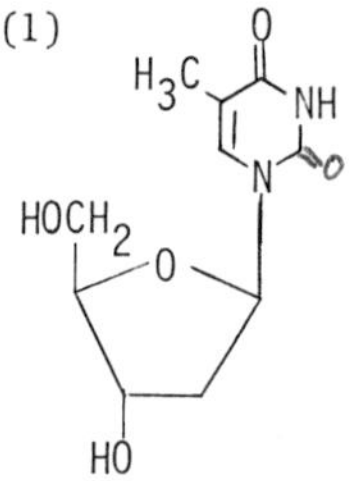

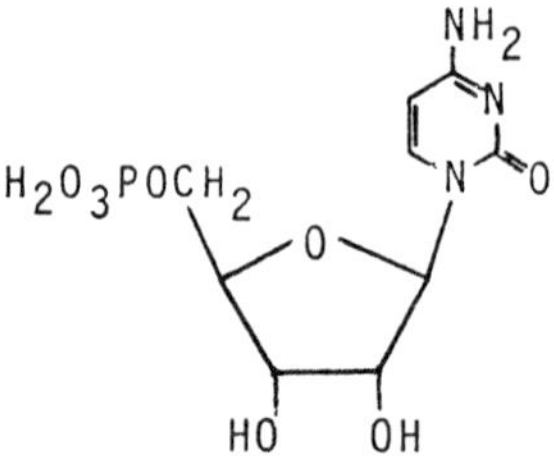

8. (1) (2) 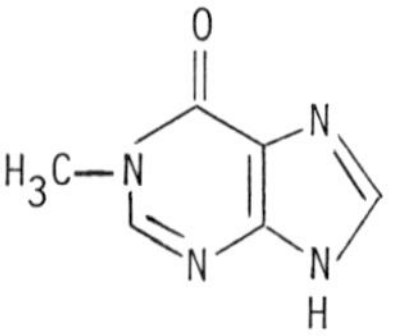 (3)

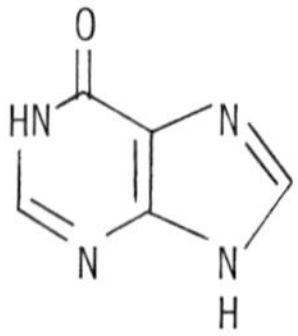

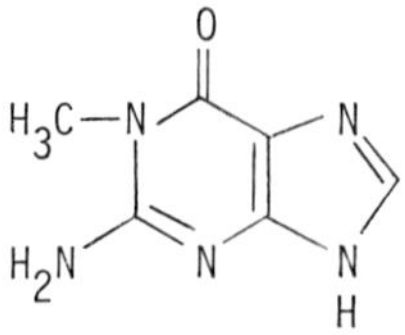

(4) 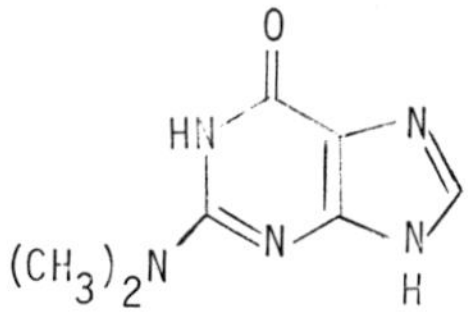 structure with $(CH_3)_2N$ substituent on purine

(5) 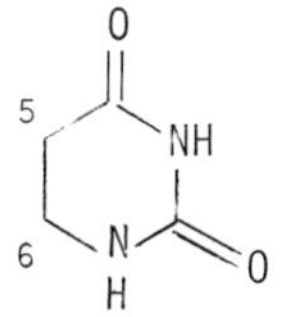 structure with note: no double bond at 5,6

(6) 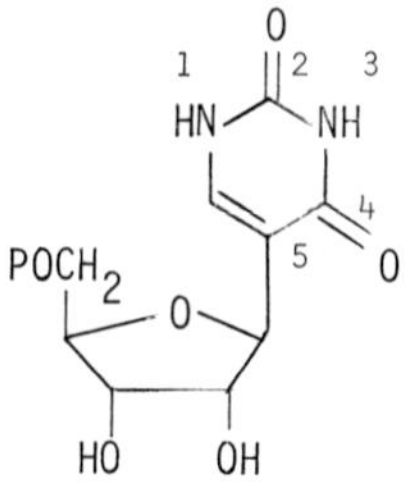 pyrimidine nucleoside with positions 1,2,3,4,5 labeled

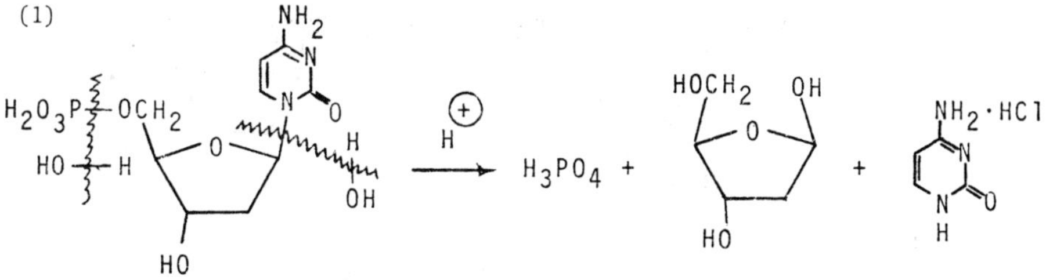

9. (1)

$$H_2O_3P\text{-}OCH_2\text{-(sugar)-cytosine} \xrightarrow{H^+} H_3PO_4 + \text{(sugar)} + \text{cytosine}\cdot HCl$$

(2)

$$H_2O_3P\text{-}OCH_2\text{-(sugar)-guanine} \xrightarrow{NaOH} \text{HOCH}_2\text{-(sugar)-guanine} + Na_3PO_4$$

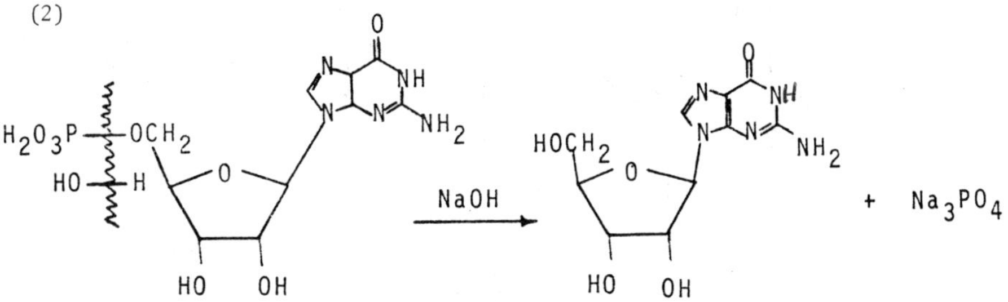

In the acidic hydrolysis in Pb. 14.9.1, the N-type glycosidic linkage is hydrolyzed; however, such a linkage is not hydrolyzed under basic conditions.

(3) [hypoxanthine tautomer with OH form shown in brackets]

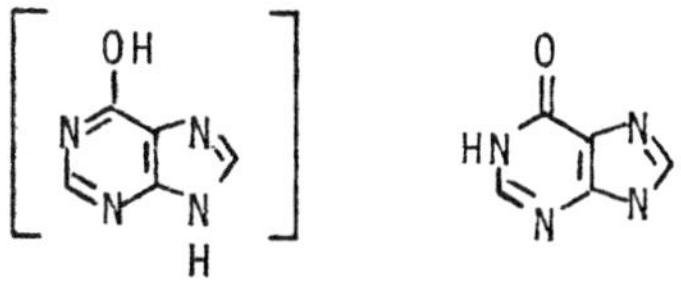

(4) hypoxanthine + AcOCH$_2$-(sugar with AcO, OAc)-N^6-acetyladenine

(5) CH$_3$OCH$_2$-(sugar with CH$_3$O, OCH$_3$)-N-methylthymine-like base

(6)

$$\begin{array}{cc} CN & CN \\ HC\text{-}OH & HO\text{-}CH \\ HC\text{-}OH & HC\text{-}OH \\ HC\text{-}OH & HC\text{-}OH \\ HC\text{-}OH & HC\text{-}OH \\ CH_2OH & CH_2OH \end{array}$$

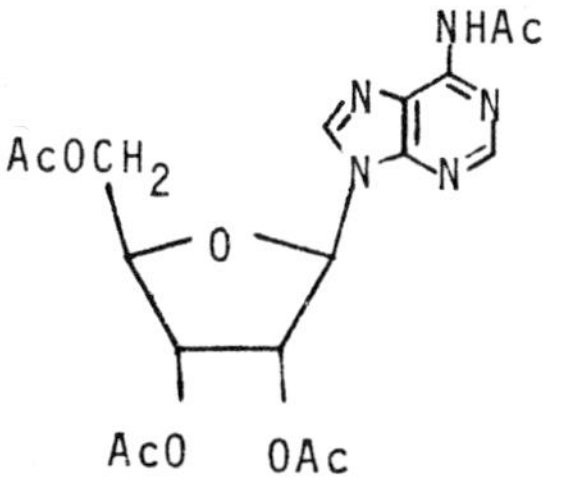

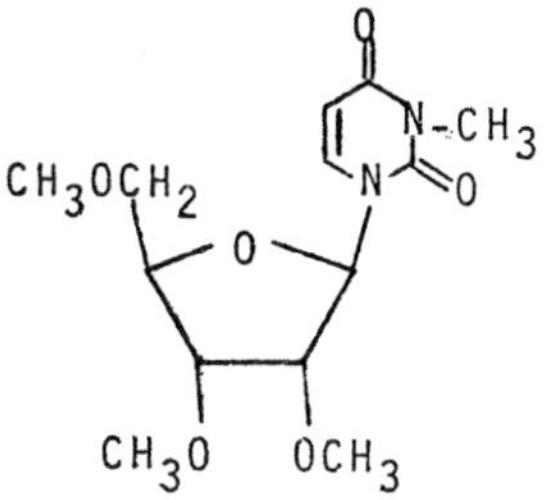

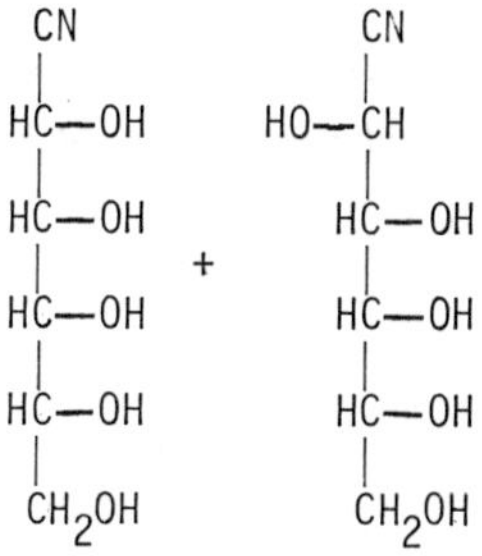

(7) 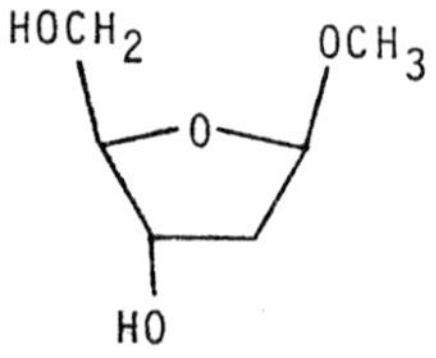 [structure: furanose ring with HOCH$_2$, OCH$_3$, HO substituents]

A second stereoisomer is formed, but one will illustrate the subsequent reactions.

(8) 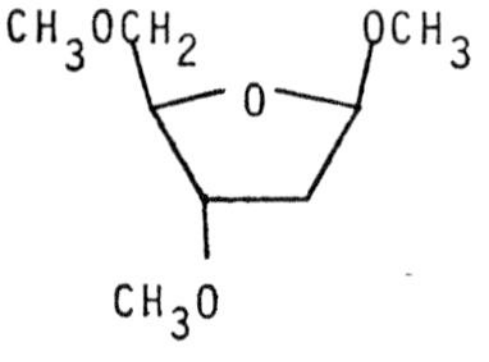 [structure: furanose with CH$_3$OCH$_2$, OCH$_3$, CH$_3$O]   (9) 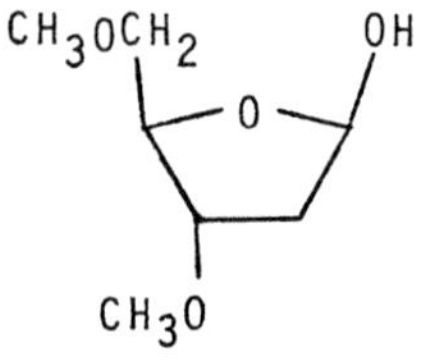 [structure: furanose with CH$_3$OCH$_2$, OH, CH$_3$O]

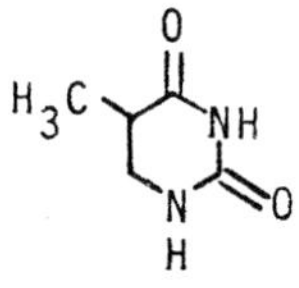

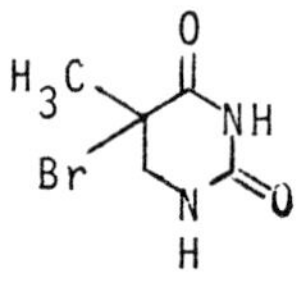

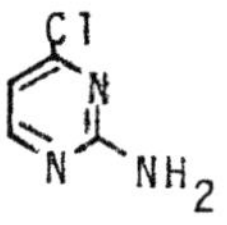

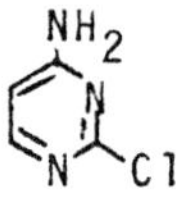

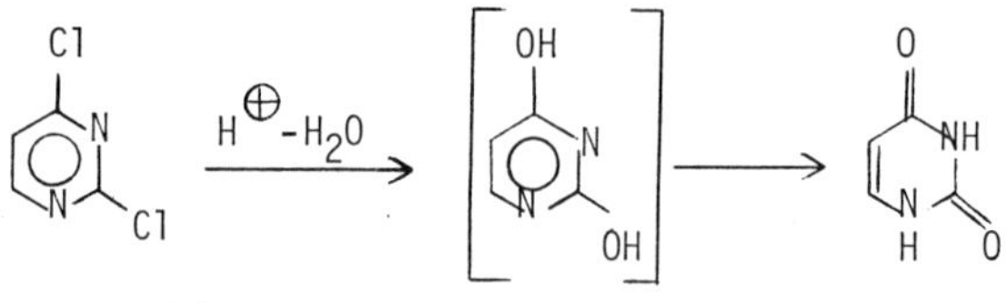

10. (1)

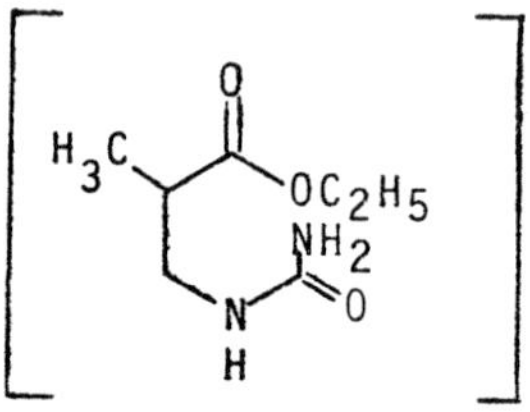

$$\left[\text{(1) H}_3\text{C-ring with OC}_2\text{H}_5\text{, NH}_2\text{, N-H, C=O} \right] \quad \text{(2) H}_3\text{C dihydropyrimidinedione} \quad \text{(3) H}_3\text{C, Br dihydropyrimidinedione}$$

(2) 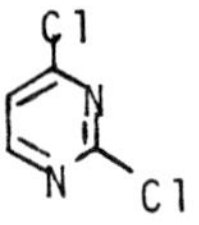
(1) 4,6-dichloropyrimidine   (2) 4-chloro-2-aminopyrimidine   (3) 4-amino-2-chloropyrimidine

[reaction scheme:]
(1) 2,4-dichloropyrimidine $\xrightarrow{H^{\oplus}, -H_2O}$ [2,4-dihydroxypyrimidine] $\longrightarrow$ uracil

Uracil can also be made from (1).

11. The cytosine units would be converted to uracil units. On replication, the units which had originally paired with guanine in the other strand would now pair with adenine nucleotide units.

chapter 15

CARBOXYLIC ACIDS AND RELATED SUBSTANCES

1.

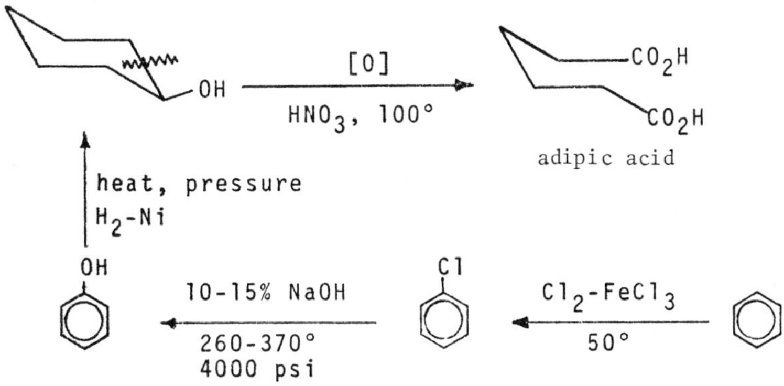

cyclohexene-OH → [O], HNO_3, 100° → adipic acid (CO_2H, CO_2H on cyclohexane)

↑ heat, pressure, H_2-Ni

phenol (OH) ← 10-15% NaOH, 260-370°, 4000 psi ← chlorobenzene (Cl) ← Cl_2-$FeCl_3$, 50° ← benzene

2.

C_6H_5-CH_3 + O_2 →[$Co(Ac)_3$-$Mn(Ac)_2$, 150-250°, pressure]→ C_6H_5-CO_2H

H_3C-C_6H_4-CH_3 →[O_2, Co^{+3}, 125-150°]→ HO_2C-C_6H_4-CO_2H

o-xylene (CH_3, CH_3) →[O_2, V_2O_5, 350-450°]→ phthalic anhydride

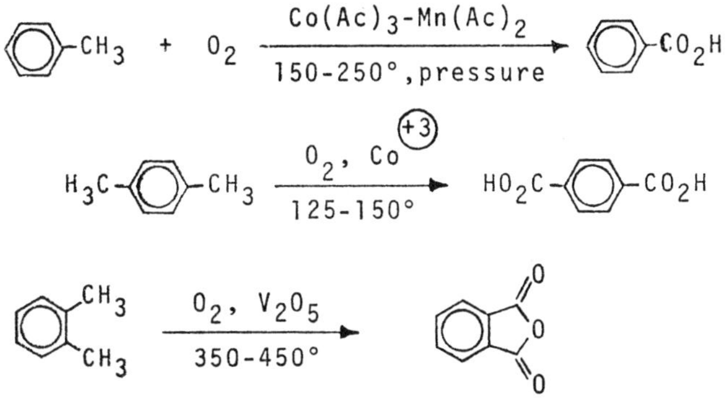

One should emphasize writing outline formulas for all rings.

3. (1) HO_2C-C_6H_3(Cl)-CO_2H

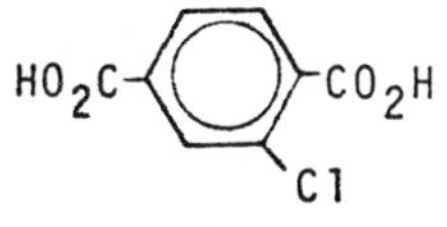

(2) HO_2C-C_6H_2(CO_2H)(Br)(CO_2H)

133

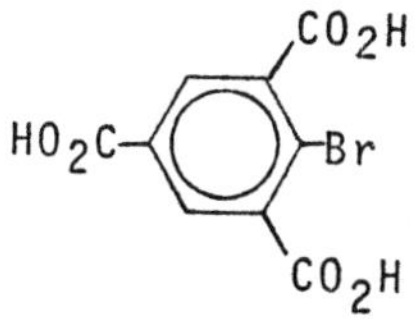

4.

$$ClCH_2CO_2H \xrightarrow{NaCN} NC-CH_2CO_2H \xrightarrow{H^{\oplus} -H_2O} HO_2C-CH_2-CO_2H$$

$$CH_2=CH_2 \xrightarrow{Br_2} BrCH_2CH_2Br \xrightarrow{NaCN} NCCH_2CH_2CN \xrightarrow{H^{\oplus} -H_2O}$$
$$HO_2C-CH_2CH_2CO_2H$$

5.

(1) $CH_3CHO \xrightarrow{HCN} CH_3\underset{OH}{CHCN} \xrightarrow{H^{\oplus} -H_2O} CH_3\underset{OH}{CHCO_2H}$

(2) $C_6H_5CHO \xrightarrow{HCN} C_6H_5\underset{OH}{CHCN} \xrightarrow{H^{\oplus} -H_2O} C_6H_5\underset{OH}{CHCO_2H}$

(3) $(CH_3)_2C=O \xrightarrow{HCN} (CH_3)_2\underset{OH}{C}-CN \xrightarrow{H^{\oplus} -H_2O} CH_2=\underset{CH_3}{C}CO_2H$

6.

$$HC\equiv CH \xrightarrow{Na} HC\equiv CNa \xrightarrow{CO_2} HC\equiv C-CO_2Na \xrightarrow{H^{\oplus} -H_2O} HC\equiv CCO_2H$$

7.

cyclopentane-$\underset{OH}{CO_2Na}$ $\xrightarrow{H^{\oplus} -H_2O}$ cyclopentane-$\underset{OH}{CO_2H}$

8.

cyclohexyl-CH_3 $\xrightarrow[(CH_3)_3COH]{HCO_2H, H_2SO_4}$ cyclohexyl-$\underset{CO_2H}{CH_3}$

9.

$$2HCO_2Na \xrightarrow{380°} NaO_2C-CO_2Na + H_2$$
$$\xrightarrow{H^{\oplus} -H_2O} HO_2C-CO_2H$$

10.

resorcinol (1,3-diOH benzene) $\xrightarrow[200°]{NH_3-NH_4Cl}$ m-aminophenol $\xrightarrow[2.\ CO_2, heat]{1.\ NaOH}$ 4-amino-2-hydroxybenzoic acid (CO_2H, OH, NH_2)

11.

3-pyridyl-CO_2H $\xrightarrow[COCl_2]{(C_2H_5)_3N}$ (3-pyridyl-C(O)-O-C(O)-3-pyridyl)

12.

$$CH_3CO_2H \xrightarrow[(CH_3CO)_2O-H_2SO_4]{Cl_2} ClCH_2CO_2H;\quad C_6H_6 \xrightarrow{Cl_2, FeCl_3, heat}$$

1,2,4,5-tetrachlorobenzene $\xrightarrow{NaOH}$ 2,4,5-trichlorophenol sodium salt $\xrightarrow{ClCH_2CO_2H}$ Cl-(2,4,5-trichlorophenyl)-O-CH$_2$CO$_2$H

13. (1) $(CH_3)_2CHCHCO_2H$ $\xrightarrow{heat}$ $(CH_3)_2CHCH_2CO_2H + CO_2$
    |
    CO_2H

(2) $CH_3COCH_2CO_2H$ $\xrightarrow{heat}$ $CH_3COCH_3 + CO_2$

(3) $O_2N\text{-}C_6H_2(NO_2)_2\text{-}CO_2H$ $\xrightarrow{heat}$ $O_2N\text{-}C_6H_3(NO_2)_2 + CO_2$

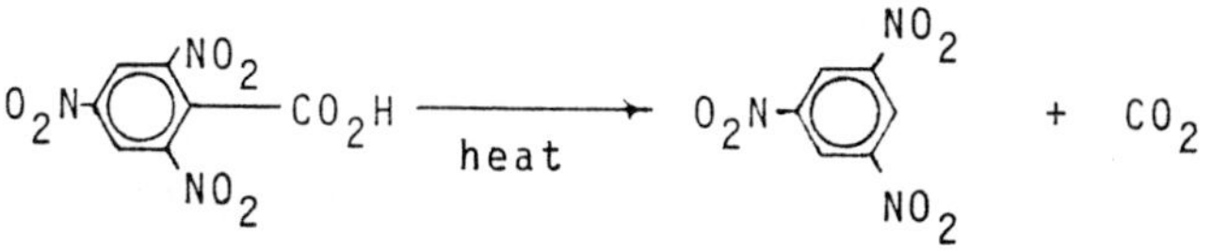

14.

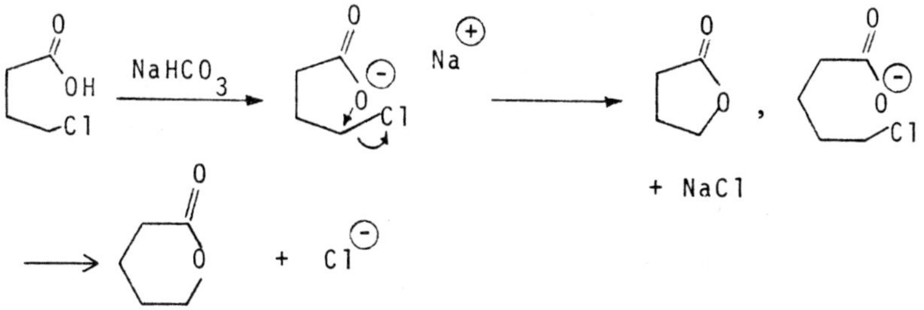

$\underset{Cl}{\overset{O}{\text{-OH}}}$ $\xrightarrow{NaHCO_3}$ (lactone intermediate with Cl) $\xrightarrow{Na^+}$ lactone, lactone $+ NaCl$

$\longrightarrow$ (6-membered lactone) $+ Cl^-$

15.

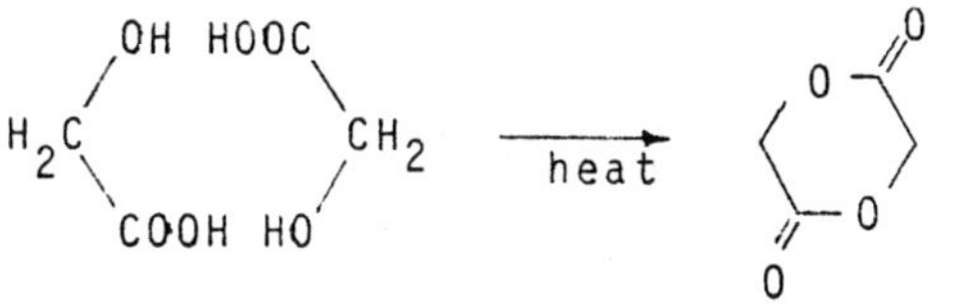

$\underset{COOH\ HO}{\overset{OH\ HOOC}{H_2C\diagdown CH_2}}$ $\xrightarrow{heat}$ a lactide $\quad CH_3\underset{OH}{CHCH_2CO_2H} \xrightarrow{heat} CH_3CH=CHCO_2H$

16.

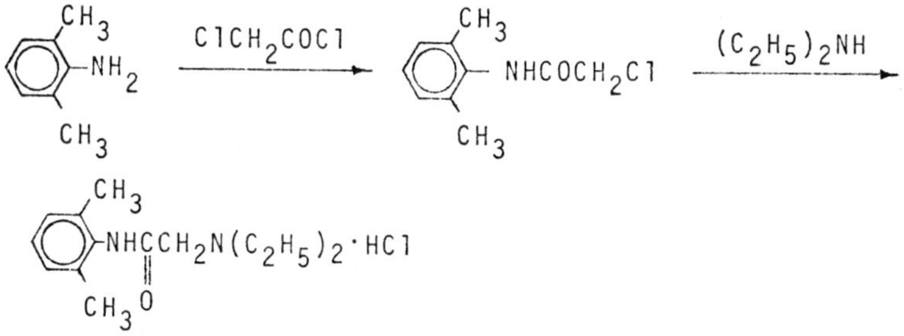

$\underset{CH_3}{\overset{CH_3}{\text{Ar-}NH_2}}$ $\xrightarrow{ClCH_2COCl}$ $\underset{CH_3}{\overset{CH_3}{\text{Ar-}NHCOCH_2Cl}}$ $\xrightarrow{(C_2H_5)_2NH}$

$\underset{CH_3}{\overset{CH_3}{\text{Ar-}NHCCH_2N(C_2H_5)_2 \cdot HCl}}$
           $\parallel$
           O

2-(diethylamino)-2',6'-dimethylacetanilide hydrochloride

17. $CO + Cl_2 \longrightarrow Cl\text{-}\underset{\parallel}{C}\text{-}Cl$
                            O

18. $\underset{CH_3CH_2CH_2\underset{CH_3}{CH}}{\overset{C_2H_5\diagup COOC_2H_5}{C\diagdown COOC_2H_5}}$ $+$ $\underset{H_2N}{\overset{H_2N}{\diagdown}}C=S$ $\xrightarrow[heat]{C_2H_5ONa}$ $\underset{CH_3CH_2CH_2\underset{CH_3}{CH}}{\overset{C_2H_5}{\diagup}} \diagdown \begin{matrix} N\text{-}Na \\ \diagup \\ = S \\ \diagdown \\ N \\ H \end{matrix}$

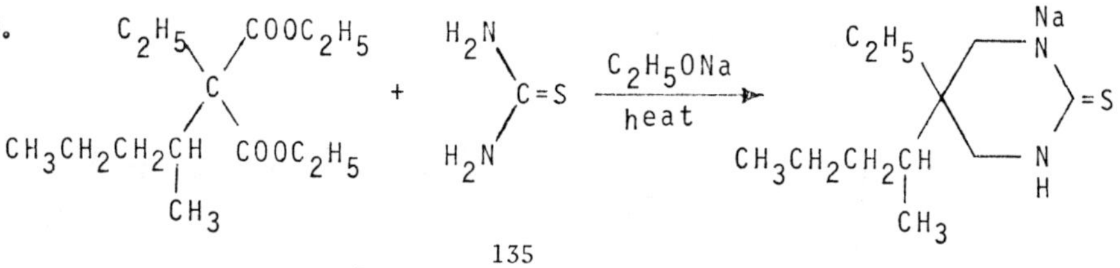

19.

$$HO_2C(CH_2)_4CO_2H \xrightarrow[\text{heat}]{NH_3} H_2NCO(CH_2)_4CONH_2 \xrightarrow[\text{heat}]{P_2O_5}$$

$$NC(CH_2)_4CN \xrightarrow[\substack{115-140° \\ \text{pressure}}]{H_2,\ Ni} H_2N(CH_2)_6NH_2$$

20.

(1) 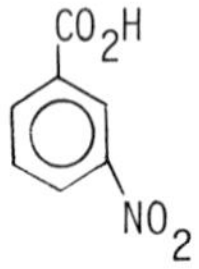 m-nitrobenzoic acid (CO$_2$H on benzene ring with NO$_2$ meta)

(2) $(CH_3)_2CHCOCl$

(3) $C_2H_5OCOCH_2CH_2COOC_2H_5$

(4) $CH_3CH_2CO_2C(CH_3)_3$

(5) 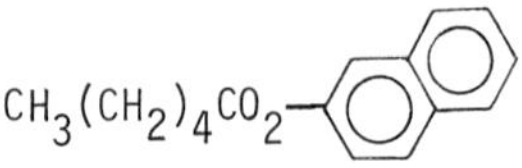 $CH_3(CH_2)_4CO_2$–(naphthyl)

(6) 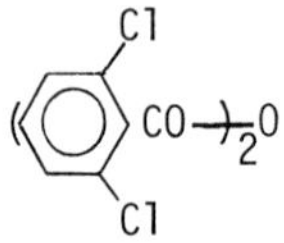 2,6-dichlorobenzoic anhydride $(CO)_2O$

(7) 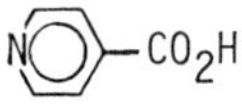 $N\hspace{-1pt}\bigcirc\hspace{-1pt}-CO_2H$

(8) $CH_3(CH_2)_3CON\begin{smallmatrix}C_2H_5\\CH_3\end{smallmatrix}$

(9) $CH_3(CH_2)_4\underset{CH(CH_3)_2}{\overset{|}{C}HCN}$

(10) $CH_3(CH_2)_6\underset{I}{\overset{CH_3}{\overset{|}{C}H}}-CH-CO_2H$

(11) $(CH_3CH_2\underset{OCH_3}{\overset{|}{C}H}-\overset{O}{\overset{\|}{C}})_2O$

(12) 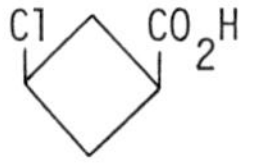 cyclobutane with Cl and CO$_2$H

(13) 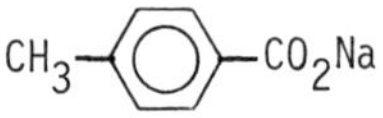 $CH_3-\bigcirc-CO_2Na$

(14) $CH_3CH_2\underset{O}{\overset{\|}{C}}(CH_2)_4CN$

21. (1) 2,5-dichlorobenzonitrile (2) benzoic anhydride
(3) m-fluorobenzamide (4) p-nitrobenzoyl chloride
(5) sodium propionate (6) tetrachlorophthalic anhydride
(7) N-phenylcycloheptanamide (8) ethyl butyrate
(9) propyl pripionate (10) m-toluic acid (11) 2-naphthoyl chloride (12) 2-methylglutaric anhydride (13) 2-cyano-2-ethylpentanoic acid (14) 3-methyl-2-butenamide

22.

(1) 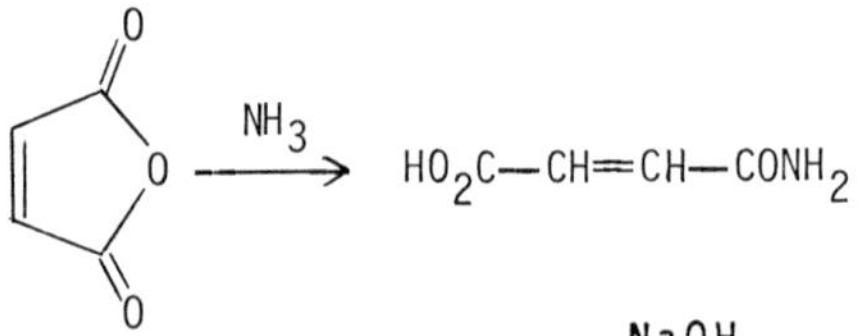 maleic anhydride $\xrightarrow{NH_3} HO_2C-CH=CH-CONH_2$

(2) $CH_3\underset{O}{\overset{\|}{C}}-O-\text{cyclohexyl} \xrightarrow{NaOH} CH_3CO_2Na + HO-\text{cyclohexyl}$

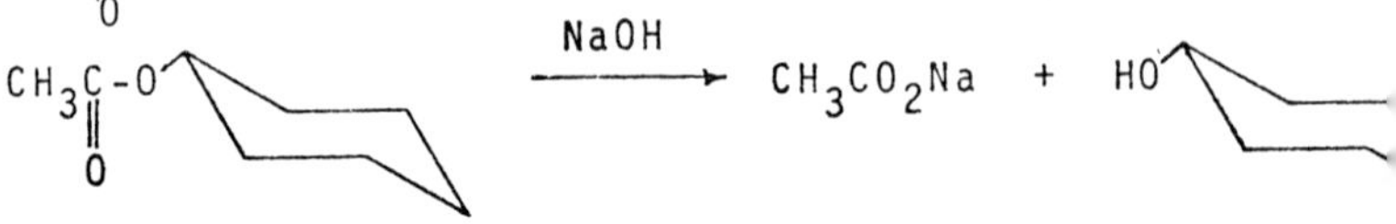

(3) [δ-valerolactone] $\xrightarrow{\text{NaOH}}$ HOCH$_2$CH$_2$CH$_2$CH$_2$CO$_2$Na

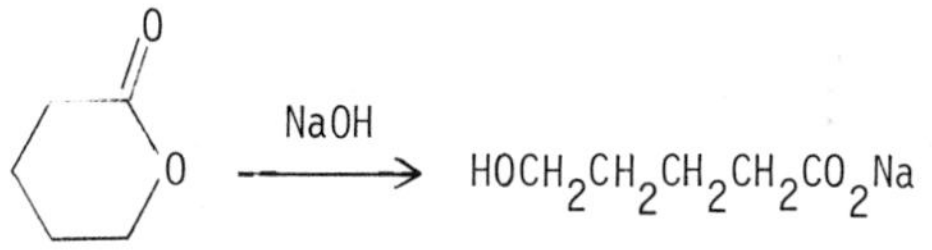

(4) CH$_3$CH$_2$CH(CH$_3$)COOH + HOCH$_2$CH$_2$CH$_3$ $\xrightarrow{H^{\oplus}}$ CH$_3$CH$_2$CH(CH$_3$)CO$_2$CH$_2$CH$_2$CH$_3$

(5) CH$_3$-C$_6$H$_4$-CH(CH$_3$)$_2$ $\xrightarrow[\text{heat}]{\text{[O]} \; KMnO_4}$ HO$_2$C-C$_6$H$_4$-CO$_2$H

(6) cyclopentyl-CO$_2$H $\xrightarrow[\text{2. H}_2\text{O}]{\text{1. Br}_2\text{-P}}$ 1-Br-cyclopentyl-CO$_2$H

    2. H$^{\oplus}$ -H$_2$O

(7) o-Cl-C$_6$H$_4$-CO$_2$K $\xrightarrow{H^{\oplus} -H_2O}$ o-Cl-C$_6$H$_4$-CO$_2$H

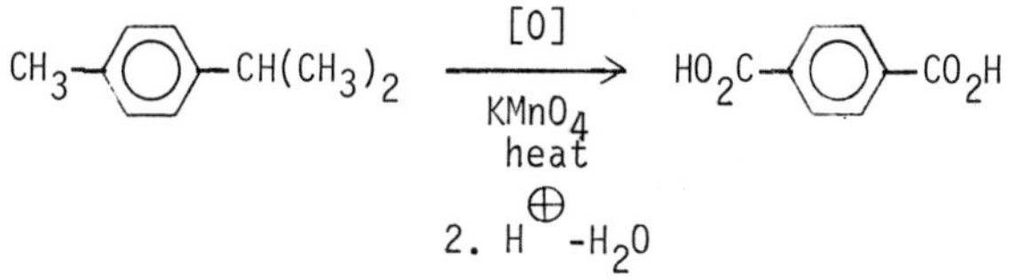

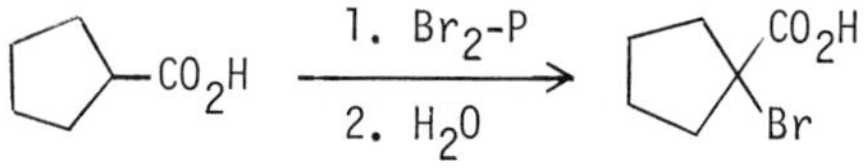

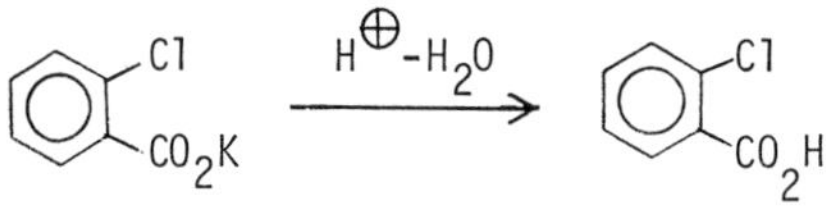

(8) CH$_3$CH$_2$CH(C$_2$H$_5$)—CH(CH$_3$)CO$_2$CH(CH$_3$)$_2$ $\xrightarrow[\text{LiAlH}_4]{\text{[H]}}$ (C$_2$H$_5$)$_2$CHCH(CH$_3$)CH$_2$OH + (CH$_3$)$_2$CHOH

(9) 1,4-(CH$_3$)$_2$C$_6$H$_4$ $\xrightarrow[\text{AlCl}_3]{\text{CH}_3\text{COCl}}$ 2,5-(CH$_3$)$_2$-C$_6$H$_3$-COCH$_3$

(10) C$_2$H$_5$OCOC$_2$H$_5$ (with C=O) $\xrightarrow[\text{K}_2\text{CO}_3]{\text{HOCH}_2\text{CH}_2\text{OH}}$ ethylene carbonate

23.

$$CH_3-\overset{\overset{\oplus}{\delta}\;O}{C}-OC_2H_5 \;+\; H_2\overset{\ominus}{\ddot{N}}NHC_6H_5 \longrightarrow CH_3-\underset{NHNHC_6H_5}{\overset{O}{C}}$$

$$CH_3-\overset{\overset{\oplus}{\delta}\;O}{C}-CH_3 \;+\; H_2\overset{\ominus}{N}NHC_6H_5 \longrightarrow CH_3-\underset{CH_3}{C}{=NNHC_6H_5}$$

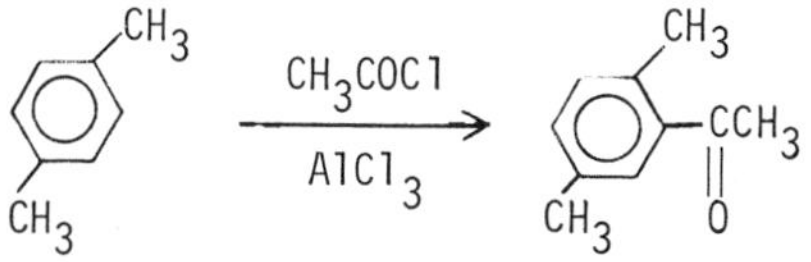

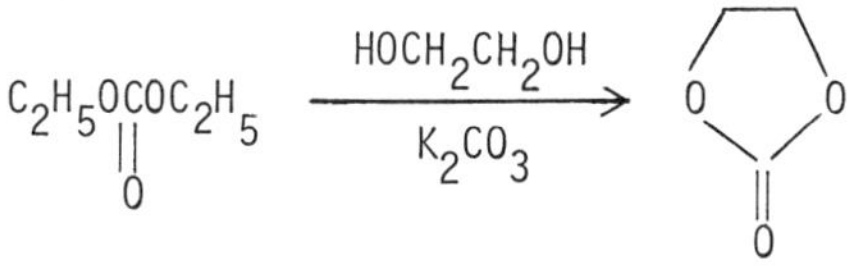

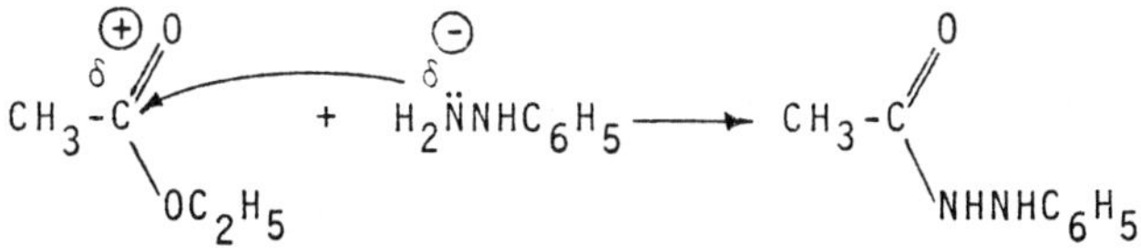

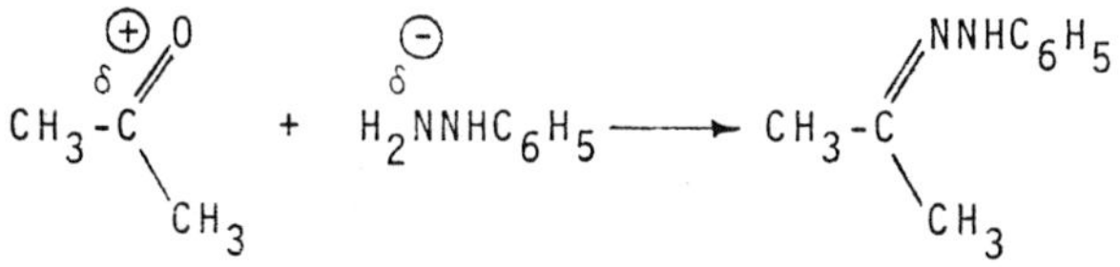

$$CH_3-\underset{\underset{OC_2H_5}{|}}{\overset{\overset{O-H}{|}}{C}}-NHNHC_6H_5 \longrightarrow CH_3-\overset{\overset{O}{\|}}{C}-NHNHC_6H_5 + C_2H_5OH$$

$$CH_3-\underset{\underset{H_3C\ H}{|}}{\overset{\overset{OH}{|}}{C}}-NNHC_6H_5 \longrightarrow CH_3-\underset{\underset{H_3C}{|}}{C}=NNHC_6H_5 + H_2O$$

The intermediate addition product is of the same type, but the subsequent elimination reaction differs.

24.

(1) $CH_3CO_2C_2H_5 \xrightarrow[\text{dry Et}_2O]{C_6H_5MgBr} CH_3\underset{\underset{C_6H_5}{|}}{\overset{\overset{C_6H_5}{|}}{C}}-OH$

(2) $(CH_3)_2\overset{\overset{\ominus\ \oplus}{\varepsilon\ \ s}}{\underset{\underset{CH_3CH_2}{|}}{C}}-MgBr\ +\ \overset{\oplus}{\underset{\underset{O}{\|}}{C}}\overset{O}{\diagup} \xrightarrow{\text{dry Et}_2O} CH_3CH_2\underset{\underset{CH_3}{|}}{\overset{\overset{CH_3}{|}}{C}}-\underset{O}{C}-OMgBr$

$\xrightarrow[-H_2O]{H^\oplus} CH_3CH_2\underset{\underset{CH_3}{|}}{\overset{\overset{CH_3}{|}}{C}}CO_2H$

(3) $C_2H_5MgBr + C_2H_5O-\overset{\overset{O}{\|}}{C}-OC_2H_5 \xrightarrow{\text{dry Et}_2O}$

$\left[C_2H_5O-\underset{\underset{O\ MgBr}{|}}{\overset{\overset{C_2H_5}{|}}{C}}-OC_2H_5 \right] \longrightarrow \left[C_2H_5O-\underset{\underset{O}{\|}}{\overset{\overset{C_2H_5}{|}}{C}} \right] \xrightarrow[\text{dry Et}_2O]{C_2H_5MgBr}$

$+\ Mg(OC_2H_5)Br$

$\left[C_2H_5O-\underset{\underset{O\ MgBr}{|}}{\overset{\overset{C_2H_5}{|}}{C}}-C_2H_5 \right] \longrightarrow \left[\underset{\underset{O}{\|}}{\overset{\overset{C_2H_5}{|}}{C}}-C_2H_5 \right] \xrightarrow[\text{dry Et}_2O]{C_2H_5MgBr}$

$C_2H_5-\underset{\underset{O\ MgBr}{|}}{\overset{\overset{C_2H_5}{|}}{C}}-C_2H_5 \xrightarrow[-H_2O]{\overset{\oplus}{H}} CH_3CH_2\underset{\underset{OH}{|}}{\overset{\overset{C_2H_5}{|}}{C}}CH_2CH_3 + \overset{\oplus}{MgBr}$

The reaction with carbonate esters is a general method for the synthesis of tertiary alcohols with three like groups. Compare this with Pb. 24.1, which is an example of the general method for the synthesis of a tertiary alcohol with two like groups.

(4) $(CH_3)_2CHCH_2\overset{\delta-}{-}MgCl$ + $\overset{\delta+}{H}-\underset{OC_2H_5}{\overset{O}{\overset{\|}{C}}}$ $\xrightarrow{\text{dry Et}_2O}$

$$\left[(CH_3)_2CHCH_2\underset{OC_2H_5}{\overset{O-MgCl}{\overset{|}{C}}}-H \right] \longrightarrow \left[(CH_3)_2CHCH_2\overset{O}{\overset{\|}{C}}-H \right]$$
isovaleraldehyde

$(CH_3)_2CHCH_2\underset{OH}{\overset{|}{C}H}CH_2CH(CH_3)_2$ ← $\begin{array}{l} (CH_3)_2CHCH_2^{\delta-}-MgCl \\ \text{dry Et}_2O \\ 2. \ \overset{+}{H}-H_2O \end{array}$

The reaction with formate esters is a general method for the synthesis of secondary alcohols with two like groups.

25. $CH_3-\underset{Cl}{\overset{O}{\overset{\|}{C}}}$ + $:NH_2$ $\longrightarrow$ $CH_3-\underset{NH_2}{\overset{O}{\overset{\|}{C}}}$ + HCl
      H

$CH_3-\underset{Cl}{\overset{O}{\overset{\|}{C}}}$ + NH_3 $\rightleftharpoons$ $CH_3-\underset{Cl}{\overset{O^-}{\overset{|}{C}}}-\overset{+}{N}H_3$ $\rightleftharpoons$

$CH_3-\underset{Cl}{\overset{:\overset{-}{O}H}{\overset{|}{C}}}-NH_2$ $CH_3-\underset{+\ Cl^-}{\overset{\overset{+}{O}H}{\overset{\|}{C}}}-NH_2$ $\xrightarrow{NH_3}$ CH_3CNH_2 + $\overset{+}{N}H_4$
$\overset{O}{\overset{\|}{}}$

26.

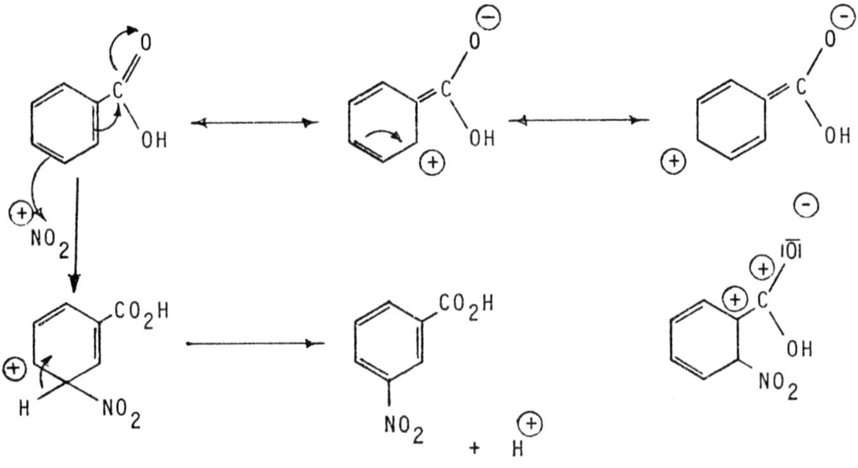

(1) [Resonance structures showing protonated benzoic acid with positive charge delocalized onto ring carbons at ortho and para positions, followed by nitronium ion attack at meta position giving meta-nitrobenzoic acid + H⁺]

(2) [Structure showing attack at ortho position placing positive charge on ring carbon bearing the electron-deficient carboxyl carbon]

Although the ring is deactivated by the electron-attracting carboxyl group as a result of general withdrawal of π electrons, the deficiency will be the least in the meta position. This is so because only the ortho and para positions can be directly affected by such resonance considerations as seen above. Thus, bond formation with the nitronium ion would take place at the meta position (1). Further, bond formation with the nitronium ion at either the ortho or para positions would place a positive charge on a ring carbon bearing the electron-deficient carboxyl carbon (2).

27.

(1) 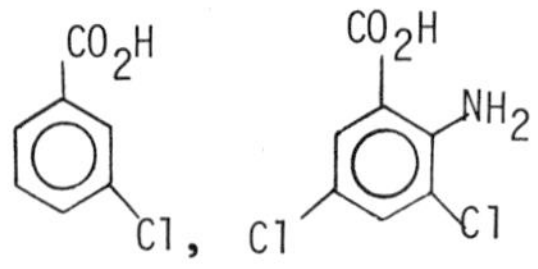 [m-chlorobenzoic acid with Cl], [2-amino-3,5-dichlorobenzoic acid]

(2) $CH_3CH_2CH_2COCl$, $CH_3CH_2CH_2CH_2Cl$

(3) 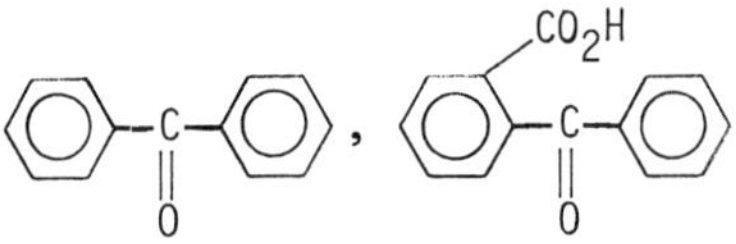 [diphenyl ketone (benzophenone)], [2-benzoylbenzoic acid structure]

(4) $CH_3CH_2CH_2NH_2$, $CH_3CH_2CH_2CH_3$

(5) $CH_3CO_2Na + CH_3OH$, $CH_3CO_2Na + CH_3NH_2$

(6) $CH_3CO_2C_2H_5$, $HO_2CCH_2CH_2CO_2C_2H_5$

28.

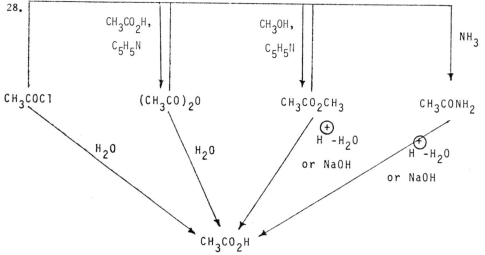

The diagram can be interpreted in the following manner. As one reads to the left, methyl acetate, acetic anhydride, and acetyl chloride each react with ammonia to give acetamide.

29.

$$CH_3C\equiv N \xrightarrow[H_2O]{H^{\oplus}} [CH_3\overset{|}{\underset{OH}{C}}=NH] \longrightarrow CH_3\underset{O}{\overset{\|}{C}}-NH_2 \xrightarrow[H_2O]{H^{\oplus}}$$

$$CH_3\underset{O}{\overset{\|}{C}}-OH + NH_4^{\oplus}$$

$$CH_3C\equiv CH \xrightarrow[\substack{H_2O \\ HgSO_4}]{H^{\oplus}} [CH_3\overset{|}{\underset{OH}{C}}=CH_2] \longrightarrow CH_3\underset{O}{\overset{\|}{C}}-CH_3$$

30. (1) $CH_3(CH_2)_6\underset{CH_3}{\overset{|}{C}}HCH_2\underset{O}{\overset{\|}{C}}-OC(CH_3)_3$ (2) [thiophene]-$CO_2^{\ominus}Na^{\oplus}$-[thiophene]-$CH_2OH$

The Cannizzaro reaction is a method of preparing carboxylic acids.

(3) $CH_3CH_2\underset{CH_3}{\overset{|}{C}}HCN$ (4) $CH_3CO_2CH_2CH_2OCOCH_3$

Reaction of more than one functional group occurs with excess acetic anhydride.

(5) $(CH_3)_2CHCH_2\underset{\underset{O}{\|}}{C}-O-\underset{\underset{O}{\|}}{C}CH_2CH(CH_3)_2$

(6) 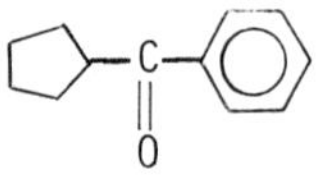 cyclopentyl–C(=O)–phenyl

(7) 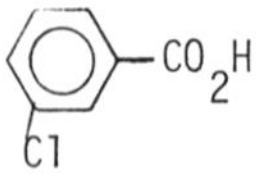 3-Cl-C$_6$H$_4$–CO$_2$H

(8) 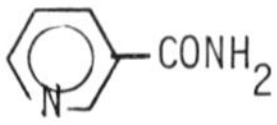 pyridyl–CONH$_2$

(9) 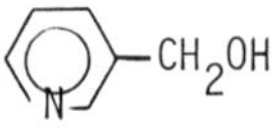 pyridyl–CH$_2$OH

(10) 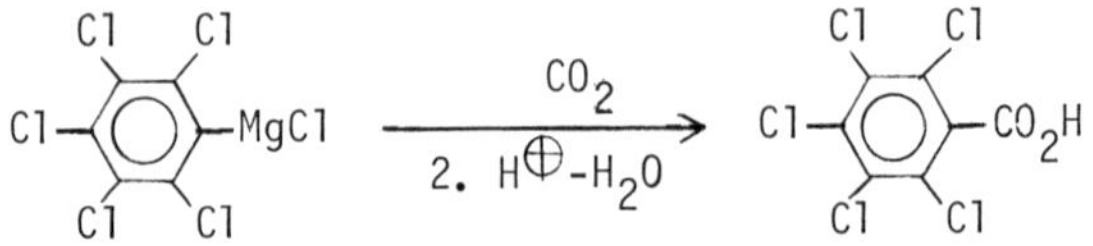 (2,3,5,6-tetrachlorophenyl)–MgCl $\xrightarrow[\text{2. } H^{\oplus}-H_2O]{CO_2}$ (2,3,5,6-tetrachlorophenyl)–CO$_2$H

This is an example of an unusual Grignard reagent, which in turn leads to an unusual acid.

31.

(1) 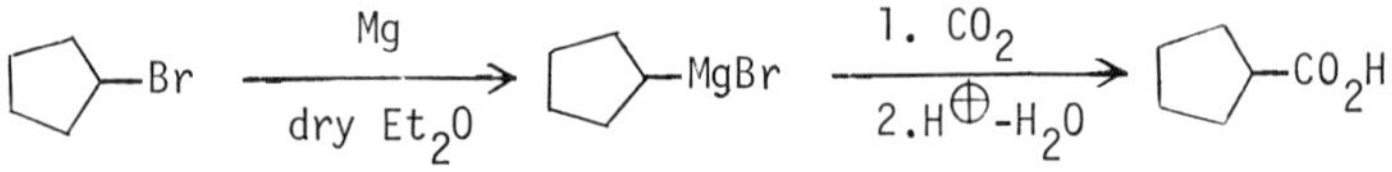 cyclopentyl–Br $\xrightarrow[\text{dry Et}_2O]{Mg}$ cyclopentyl–MgBr $\xrightarrow[\text{2. }H^{\oplus}-H_2O]{\text{1. }CO_2}$ cyclopentyl–CO$_2$H

(2) 2,6-dimethylphenyl–Br $\xrightarrow[\text{dry Et}_2O]{Mg}$ 2,6-dimethylphenyl–MgBr $\xrightarrow[\text{2. }H^{\oplus}-H_2O]{\text{1. }CO_2}$ 2,6-dimethylphenyl–CO$_2$H

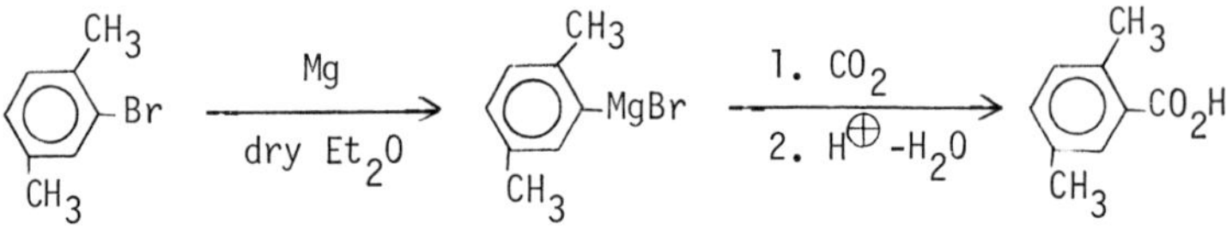

(3) 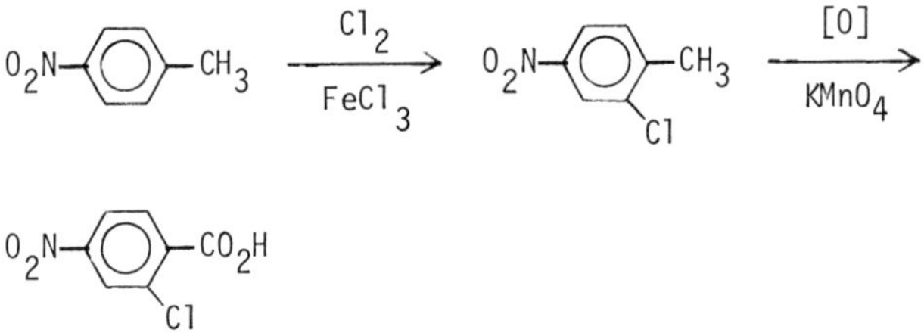 O_2N–C$_6$H$_4$–CH$_3$ $\xrightarrow[FeCl_3]{Cl_2}$ O_2N–(3-Cl-4-CH$_3$-phenyl) $\xrightarrow[KMnO_4]{[O]}$

O_2N–(3-Cl-phenyl)–CO$_2$H

(4) $CH_3(CH_2)_4\underset{F}{\overset{|}{C}}HCH_2OCOCH_3$ $\xrightarrow[\text{[O]}]{HNO_3}$ $CH_3(CH_2)_4\underset{F}{\overset{|}{C}}HCO_2H$

$\xrightarrow[\text{2. }NH_3]{\text{1. }SOCl_2}$ $CH_3(CH_2)_4\underset{F}{\overset{|}{C}}HCONH_2$

Hydrolysis of ester occurs first, followed by oxidation of the 1° alcohol group to —CO$_2$H.

(5) $CH_3(CH_2)_3\underset{CH_3}{\overset{|}{C}}HCO_2H$ $\xrightarrow[\text{2. }C_6H_5NH_2]{\text{1. }SOCl_2}$ $CH_3(CH_2)_3\underset{CH_3}{\overset{|}{C}}HCONHC_6H_5$

(6) $C_6H_5CO_2H \xrightarrow{SOCl_2} C_6H_5COCl \xrightarrow[\text{pyridine}]{C_6H_5OH} C_6H_5CO_2C_6H_5$

(7) 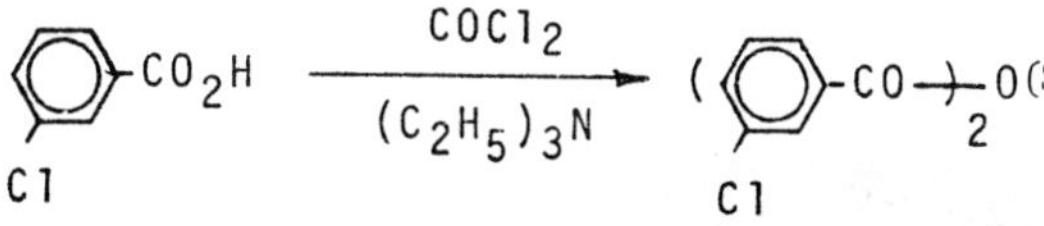 $\text{4-Cl-}C_6H_4\text{-}CO_2H \xrightarrow[(C_2H_5)_3N]{COCl_2} (\text{4-Cl-}C_6H_4\text{-}CO)_2O$ (See Pb. 15.11.)

(8) $(CH_3)_3CCl \xrightarrow[\substack{\text{2. } CO_2^{\oplus} \\ \text{3. } H^+\text{-}H_2O}]{\text{1. Mg, dry Et}_2O} (CH_3)_3C\text{-}CO_2H \xrightarrow{LiAlH_4}$

$(CH_3)_3C\text{-}CH_2OH$

chapter 16

GLYCERIDES: FATS AND OILS

1. 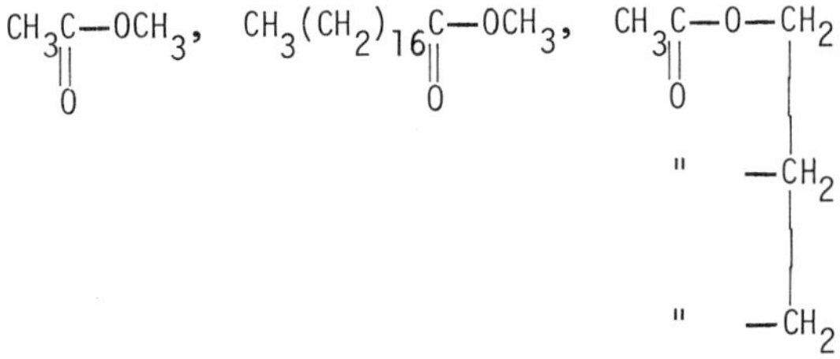 $CH_3\overset{O}{\underset{\parallel}{C}}-OCH_3$, $CH_3(CH_2)_{16}\overset{O}{\underset{\parallel}{C}}-OCH_3$, $CH_3\overset{O}{\underset{\parallel}{C}}-O-CH_2$

   $" \quad -CH_2$

   $" \quad -CH_2$

   Ditto marks may be used in glyceride formulas where the acid residues are identical.

2. 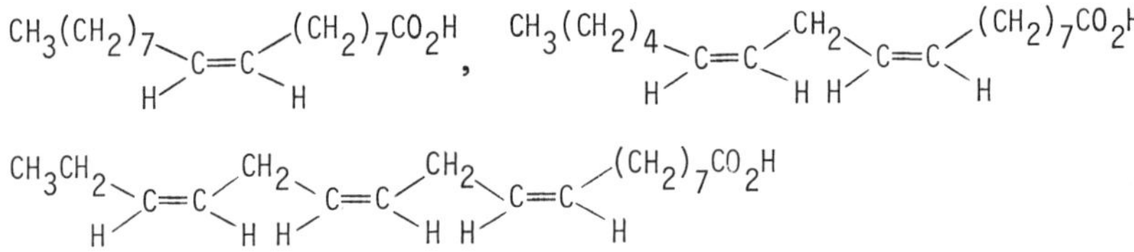 $CH_3(CH_2)_7\underset{H}{\overset{}{C}}=\underset{H}{\overset{(CH_2)_7CO_2H}{C}}$, $CH_3(CH_2)_4\underset{H}{\overset{}{C}}=\underset{H}{\overset{CH_2}{C}}\underset{H}{\overset{}{C}}=\underset{H}{\overset{(CH_2)_7CO_2H}{C}}$

   $CH_3CH_2\underset{H}{\overset{}{C}}=\underset{H}{\overset{CH_2}{C}}\underset{H}{\overset{}{C}}=\underset{H}{\overset{CH_2}{C}}\underset{H}{\overset{}{C}}=\underset{H}{\overset{(CH_2)_7CO_2H}{C}}$

3. $CH_3(CH_2)_7\underset{CH_3}{\overset{}{CH}}(CH_2)_8CO_2H$, ⬠$-(CH_2)_{12}CO_2H$

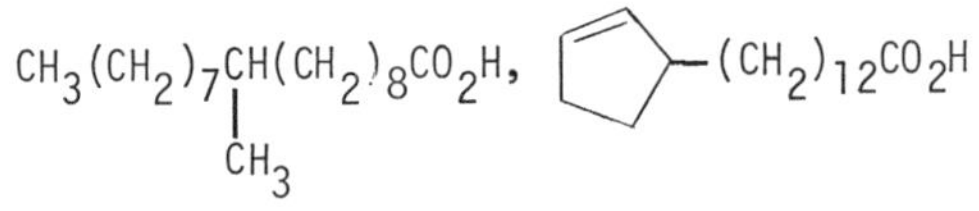

   Note that chaulmoogric acid, like stearic and oleic acids, contains 18 C atoms.

4. The number of equivalents of KOH (MW 56) reacting with 1 g of fat is given by $\frac{240}{3 \times 56}$ (in the equation three equivalents of KOH are required). This is equal to the weight of fat (1000 mg) divided by the unknown average molecular weight (x). Thus, $\frac{1000}{x} = \frac{240}{3 \times 56}$; x = 700. To obtain the average weight of the carboxylic acid

residues, one needs to subtract the weight of $-O-CH_2$ from the average molecular weight; thus, $C_3H_5O_3 = 89$, and we have $700 - 89 = 611$. Since there are three carboxylic acid residues in a glyceride, we obtain an average weight of $\frac{611}{3} = \sim 204$ per residue.

$$\begin{array}{l} -O-CH_2 \\ | \\ -O-CH \\ | \\ -O-CH_2 \end{array}$$

5. Equivalents of glycerides are given by the ratio $\frac{100}{950}$, as 100 g of fat of average molecular weight of 950 are reacting with 187 g of iodine ($I_2 = 254$). Thus, the equivalents of iodine reacting with the fat are given by the expression $\frac{187}{254d}$ where d is the number of double bonds per glyceride molecule. Thus, $\frac{100}{950} = \frac{187}{254d}$, and $d = \frac{950 \times 187}{25400} = 7$.

6. Coconut oil, which yields 44-46% lauric acid, would be the most likely fat or oil for the production of lauryl alcohol.

$$\begin{array}{c} CH_3(CH_2)_{10}\overset{O}{\underset{\|}{C}}-OCH_2 \\ \phantom{CH_3(CH_2)_{10}C}| \\ "\phantom{CH_3(CH_2)_{10}C}-OCH \\ \phantom{CH_3(CH_2)_{10}C}| \\ "\phantom{CH_3(CH_2)_{10}C}-OCH_2 \end{array} \xrightarrow[3000-4500 \text{ psi}]{\underset{CuO-CuCr_2O_4}{H_2}} 3\ CH_3(CH_2)_{10}CH_2OH\ +\ \begin{array}{c} HOCH_2 \\ | \\ HOCH \\ | \\ HOCH_2 \end{array}$$

7. 1-Hexadecanol is obtained by hydrogenation of an appropriate fat or oil, e.g., tallow.

$$CH_3(CH_2)_{14}CH_2OH \xrightarrow{HCl} CH_3(CH_2)_{14}CH_2Cl \xrightarrow[\text{heat, pressure}]{NH_3}$$

$$CH_3(CH_2)_{14}CH_2NH_2 \xrightarrow{3\ CH_3Cl} CH_3(CH_2)_{14}CH_2\overset{\oplus}{N}(CH_3)_3Cl^{\ominus}$$

$$CH_3(CH_2)_8Cl\ +\ \bigcirc\!\!-OH \xrightarrow{AlCl_3} CH_3(CH_2)_8\!\!-\!\!\bigcirc\!\!-OH$$

$$\xrightarrow{\overset{O}{\triangle}} CH_3(CH_2)_8\!\!-\!\!\bigcirc\!\!-O(CH_2CH_2)_{6-122}CH_2CH_2OH$$

This polymerization reaction with ethylene oxide is new.

8. The process is illustrated by using just the conjugate segment of eleostearic acid.

$$\sim\sim CH=CH-CH=CH-CH=CH \sim\sim + R\cdot \longrightarrow$$

$$\sim\sim CH-\underset{R}{\overset{|}{CH}}-CH=CH-CH=CH \sim\sim \quad \xrightarrow{O_2}$$

$$\sim\sim CH-\underset{R}{\overset{|}{CH}}-\underset{O-O\cdot}{\overset{|}{CH}}=CH-CH=CH \sim\sim \quad \xrightarrow{\sim\sim CH=CH-CH=CH-CH=CH\sim\sim}$$

$$\sim\sim CH-\underset{R}{\overset{|}{CH}}-\underset{O-O}{\overset{|}{CH}}=CH-CH=CH \sim\sim$$
$$\quad\quad\quad\quad\quad\quad\quad |$$
$$\sim\sim CH-CH=CH-CH=CH-CH \sim\sim$$

1. O_2
2. Further

$\xrightarrow{\text{eleostearic residue}}$

$$\sim\sim CH-\underset{R}{\overset{|}{CH}}-\underset{O-O}{\overset{|}{CH}}=CH-CH=CH \sim\sim$$
$$\quad\quad\quad\quad |$$
$$\sim\sim CH-CH=CH-CH=CH-CH \sim\sim$$
$$\quad\quad\quad\quad\quad\quad\quad\quad\quad\quad |$$
$$\quad\quad\quad\quad\quad\quad\quad\quad\quad\quad O-O$$
$$\quad\quad\quad\quad\quad\quad\quad\quad\quad\quad\quad |$$
$$\sim\sim CH=CH-CH=CH-CH-CH \sim\sim$$

In linseed oil, free radical attack occurs at the methylene groups separating double bonds. The resulting free radicals then isomerize to give conjugate systems. Rancidity in fats and oils also is partially a result of such autooxidation reactions; a mixture of aldehydes, ketones, and acids results and imparts an unsavory flavor or odor to the food product in question.

9.

$$CH_3(CH_2)_8\overset{O}{\overset{\|}{C}}OCH_2$$
$$CH_3(CH_2)_2\overset{O}{\overset{\|}{C}}OCH$$
$$CH_3(CH_2)_8\overset{O}{\overset{\|}{C}}OCH_2$$

$\xrightarrow[H^{\oplus}]{H_2O}$

$CH_3(CH_2)_2CO_2H$
butyric acid

+

$CH_3(CH_2)_8CO_2H$
capric acid

+

$HOCH_2$
$|$
$HOCH$
$|$
$HOCH_2$

glycerol 2-butyrate 1,3-didecanoate

10. $CH_3(CH_2)_{14}\overset{O}{\underset{\|}{C}}-O(CH_2)_{15}CH_3 \xrightarrow[\text{3000-4500 psi}]{\overset{H_2}{\text{CuO—CuCr}_2O_4}} CH_3(CH_2)_{15}OH$

$\xrightarrow{H_2SO_4} CH_3(CH_2)_{15}OSO_2OH \xrightarrow{NaOH} CH_3(CH_2)_{15}OSO_2O^{\ominus} \; Na^{\oplus}$

11. $CH_3(CH_2)_{14}\overset{O}{\underset{\|}{C}}CH_2\overset{O}{\underset{\|}{C}}-SCoA + H\ddot{S}-CoA \longrightarrow$

$CH_3(CH_2)_{14}\underset{O-H}{\overset{SCoA}{\underset{|}{C}}}-CH_2\overset{O}{\underset{\|}{C}}-SCoA \longrightarrow CH_3(CH_2)_{14}\overset{SCoA}{\underset{\underset{\|}{O}}{\underset{|}{C}}} + CH_3\overset{O}{\underset{\|}{C}}-SCoA$

12. $\sim\sim\sim CH_2\overset{O}{\underset{\|}{C}}CH_2\overset{O}{\underset{\|}{C}}-SCoA \longrightarrow \sim\sim\sim CH_2\underset{OH}{\underset{|}{CH}}CH_2COSCoA$

$\longrightarrow \sim\sim\sim CH_2CH=CHCOSCoA \longrightarrow \sim\sim\sim CH_2CH_2CH_2COSCoA$

13.
$CH_3(CH_2)_7CH=CH(CH_2)_7COSCoA + \begin{array}{c} HOCH_2 \\ | \\ HOCH \\ | \\ HOPO_2OCH_2 \end{array} \longrightarrow$

$CH_3(CH_2)_7CH=CH(CH_2)_7\overset{O}{\underset{\|}{C}}-O\underset{\begin{array}{c}|\\HOPO_2OCH_2\end{array}}{\overset{\begin{array}{c}HOCH_2\\|\end{array}}{CH}} \xrightarrow{CH_3(CH_2)_{14}COSCoA}$

monoglyceride phosphate

+ HS-CoA

$CH_3(CH_2)_7CH=CH(CH_2)_7\overset{O}{\underset{\|}{C}}-O\underset{\begin{array}{c}|\\HOPO_2OCH_2\end{array}}{\overset{\begin{array}{c}CH_3(CH_2)_{14}\overset{O}{\underset{\|}{C}}-OCH_2\\|\end{array}}{CH}} \xrightarrow{\begin{array}{l}\text{1. phosphatase}\\\text{2. }CH_3(CH_2)_{16}COSCoA\end{array}}$

diglyceride phosphate

+ HS-CoA

$$\begin{array}{c}
CH_3(CH_2)_{14}\underset{\underset{O}{\|}}{C}-OCH_2 \\
CH_3(CH_2)_7CH=CH(CH_2)_7\underset{\underset{O}{\|}}{C}-OCH \\
CH_3(CH_2)_{16}\underset{\underset{O}{\|}}{C}-OCH_2
\end{array} \quad \begin{array}{l} + H_3PO_4 \\ \\ + HS\text{-}CoA \end{array}$$

triglyceride

14. (1) $CH_3(CH_2)_3CH=CH-CH=CH-CH=CH(CH_2)_7\underset{\underset{O}{\|}}{C}-OCH_2$ less-preferred name—glycerol trieleostearate

$\qquad\qquad\qquad\qquad\qquad\qquad\qquad\qquad\quad\;\; "\qquad\qquad\qquad -OCH$

$\qquad\qquad\qquad\qquad\qquad\qquad\qquad\qquad\quad\;\; "\qquad\qquad\qquad -OCH_2$

(2) $CH_3(CH_2)_8\underset{\underset{O}{\|}}{C}-OCH_2$ \qquad (3) $\quad CH_3(CH_2)_{12}\underset{\underset{O}{\|}}{C}-OCH_2$

$\qquad\quad\;\; " \qquad\; -OCH \qquad\qquad\qquad CH_3(CH_2)_7CH=CH(CH_2)_7\underset{\underset{O}{\|}}{C}-OCH$

$\qquad\quad\;\; " \qquad\; -OCH_2 \qquad\qquad\qquad\qquad\qquad\qquad CH_3(CH_2)_{12}\underset{\underset{O}{\|}}{C}-OCH_2$

(4) $CH_3(CH_2)_4CH=CHCH_2CH=CH(CH_2)_7\underset{\underset{O}{\|}}{C}-OCH_2$

$\quad\; CH_3(CH_2)_5\underset{OH}{\overset{}{C}H}CH_2CH=CH(CH_2)_7\underset{\underset{O}{\|}}{C}-OCH$

$\qquad\qquad\qquad\qquad " \qquad\qquad\qquad\qquad -OCH_2$

15. (1) $[CH_3(CH_2)_{16}CO_2]_2Ca$ \qquad (2) $[CH_3(CH_2)_{10}CO]_2O$

(3) $CH_3(CH_2)_{15}OH$

(4) $CH_3CH_2CH=CHCH_2CH=CHCH_2CH=CH(CH_2)_7\underset{\underset{O}{\|}}{C}-O(CH_2)_8CH=CHCH_2CH=CH(CH_2)_4CH_3$

(5) $CH_3(CH_2)_7CH=CH(CH_2)_7\underset{\underset{O}{\|}}{C}-Cl$

(6) $CH_3(CH_2)_{12}\underset{\underset{O}{\|}}{C}-NH_2$

16.

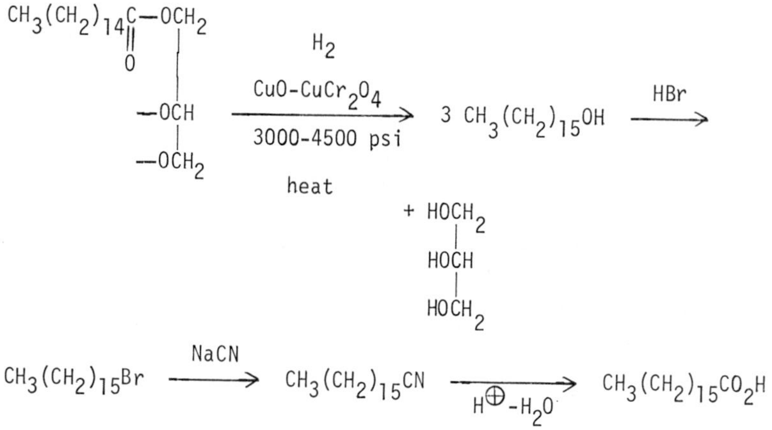

$$CH_3(CH_2)_{14}\overset{O}{\underset{|}{C}}-OCH_2$$
$$\phantom{CH_3(CH_2)_{14}C}-OCH$$
$$\phantom{CH_3(CH_2)_{14}C}-OCH_2$$

$\xrightarrow[\text{heat}]{\underset{\text{3000-4500 psi}}{\text{CuO-CuCr}_2\text{O}_4}}^{H_2}$ $3\ CH_3(CH_2)_{15}OH \xrightarrow{HBr}$

$+\ HOCH_2-HOCH-HOCH_2$

$CH_3(CH_2)_{15}Br \xrightarrow{NaCN} CH_3(CH_2)_{15}CN \xrightarrow[H^{\oplus}-H_2O]{} CH_3(CH_2)_{15}CO_2H$

17.

(1) $C_6H_5\overset{O}{\underset{\|}{C}}-Cl\ +\ HOCH_2-HOCH-HOCH_2 \xrightarrow{C_5H_5N} C_6H_5\overset{O}{\underset{\|}{C}}-OCH_2$

   " —OCH

   " —OCH$_2$

(2) $CH_3(CH_2)_{10}\overset{O}{\underset{\|}{C}}-OCH_2$

   " —OCH

   " —OCH$_2$

   $\xrightarrow{NH_3}\ CH_3(CH_2)_{10}\overset{O}{\underset{\|}{C}}-NH_2\ +\ HOCH_2-HOCH-HOCH_2$

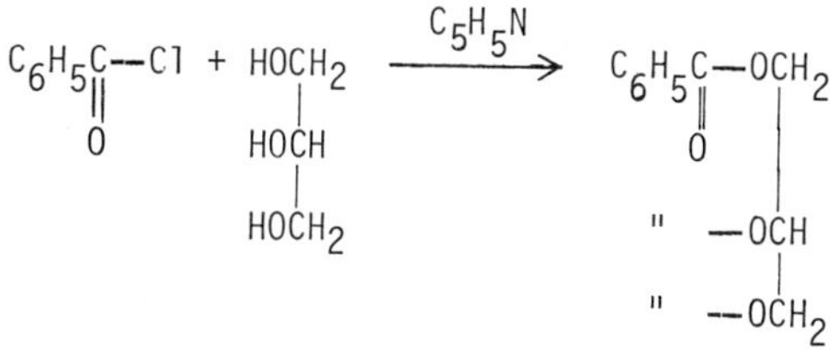

(3) $CH_3(CH_2)_7CH=CH(CH_2)_7\overset{O}{\underset{\|}{C}}-OCH_2$

   " —OCH

   " —OCH$_2$

   $\xrightarrow[\text{heat}]{\underset{\text{3000-4500 psi}}{\text{CuO-CuCr}_2\text{O}_4}}^{H_2}$

   $CH_3(CH_2)_{16}CH_2OH\ +\ HOCH_2-HOCH-HOCH_2$

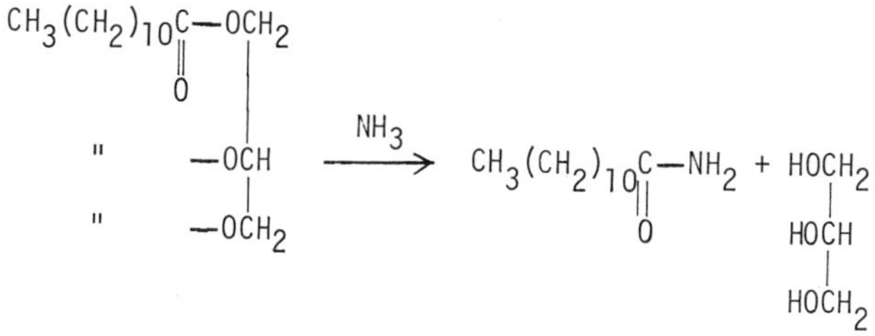

(4) $CH_3(CH_2)_4\overset{O}{\underset{\|}{C}}-OCH_2$

   " —OCH

   " —OCH$_2$

   $\xrightarrow[\text{[H]}]{LiAlH_4}\ CH_3(CH_2)_5OH\ +\ HOCH_2-HOCH-HOCH_2$

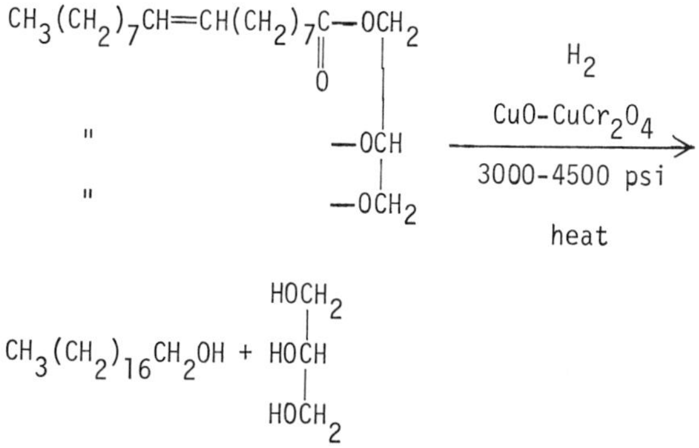

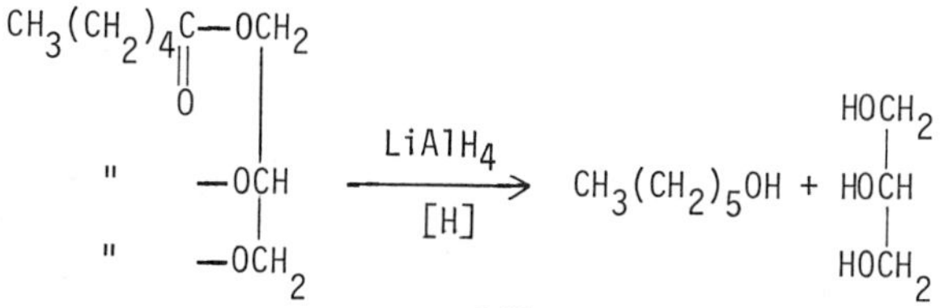

18. (1) This resolves itself into a stoichiometric calculation on the basis of the number of milligrams of KOH required for the saponification of 1 gram of fat; saponification number (SN).

$$\frac{1}{806} = \frac{SN}{3 \times 56} \qquad SN = \frac{168}{806} = .208 \text{ g or } 208 \text{ mg KOH}$$

(2) $\frac{1}{884} = \frac{SN}{3 \times 56} \qquad SN = \frac{168}{884} = .190 \text{ or } 190 \text{ mg}$

19. (1) Since the iodine number (IN) is the number of grams of iodine which will react with 100 g of fat or oil, then:

$$\frac{100}{932} = \frac{IN}{3 \times 56} \qquad IN = \frac{76200}{932} = 82$$

(2) $\frac{100}{872} = \frac{IN}{9 \times 254} \qquad IN = \frac{228600}{872} = 262$

chapter 17
PEPTIDES AND PROTEINS (AMINO ACID POLYMERS)

1.
(1) 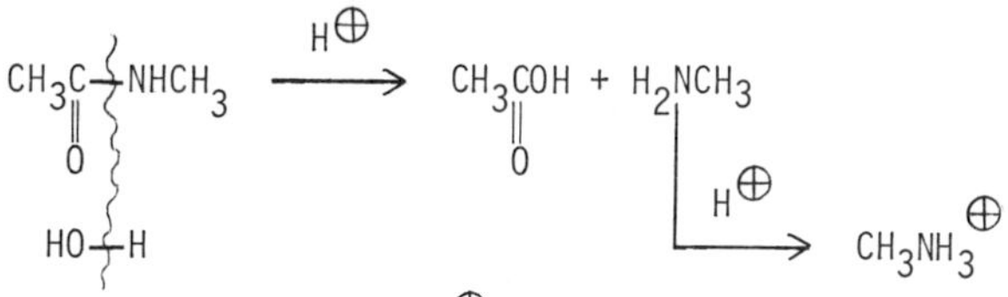 $CH_3C(=O)-NHCH_3 \xrightarrow{H^\oplus} CH_3COH + H_2NCH_3$
   $HO-H$
   $\xrightarrow{H^\oplus} CH_3NH_3^\oplus$

(2) 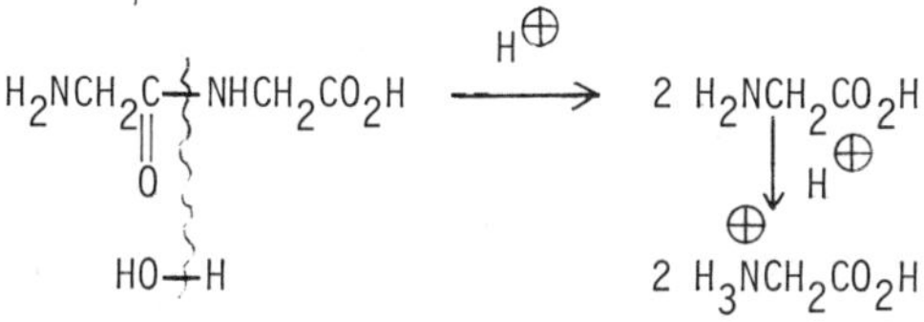 $H_2NCH_2C(=O)-NHCH_2CO_2H \xrightarrow{H^\oplus} 2\ H_2NCH_2CO_2H$
   $HO-H$
   $\downarrow H^\oplus$
   $2\ H_3^\oplus NCH_2CO_2H$

2.
(1) 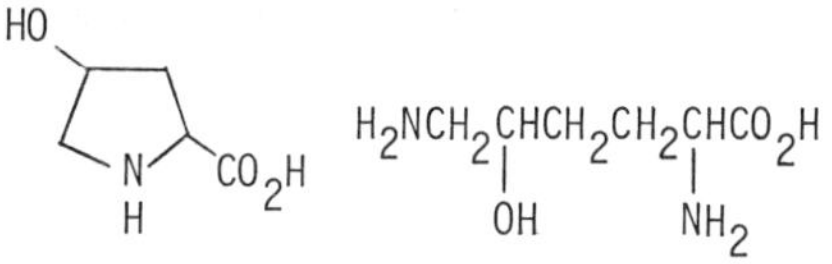 HO-[pyrrolidine]-CO_2H (N-H)

   $H_2NCH_2CHCH_2CH_2CHCO_2H$
   $\quad\quad\quad |\quad\quad\quad\quad |$
   $\quad\quad\quad OH\quad\quad\quad NH_2$

(2) $HO-$[C$_6$H$_2$I$_2$]$-O-$[C$_6$H$_2$I$_2$]$-CH_2CHCO_2H$
   $\quad\quad\quad\quad\quad\quad\quad\quad\quad\quad\quad\quad |$
   $\quad\quad\quad\quad\quad\quad\quad\quad\quad\quad\quad\quad NH_2$

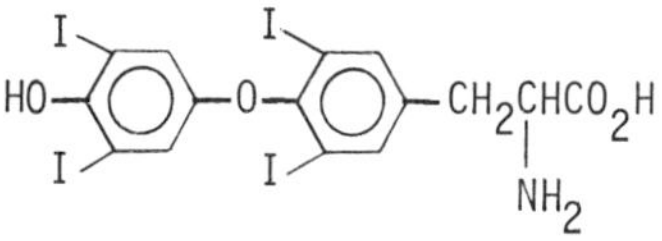

(3) 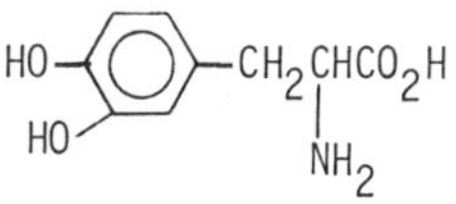 $HO-$[C$_6$H$_3$(OH)]$-CH_2CHCO_2H$
   $\quad\quad\quad\quad\quad\quad\quad\quad |$
   $\quad\quad\quad\quad\quad\quad\quad\quad NH_2$

3.
(1) $(CH_3)_2CHCH_2CH_2CO_2H \xrightarrow[2.\ H^\oplus\ -H_2O]{1.\ P-Br_2} (CH_3)_2CHCH_2CHCO_2H$
   $\quad\quad\quad\quad\quad\quad\quad\quad\quad\quad\quad\quad\quad\quad\quad\quad\quad\quad |$
   $\quad\quad\quad\quad\quad\quad\quad\quad\quad\quad\quad\quad\quad\quad\quad\quad\quad\quad Br$

(2) $(CH_3)_2CHCH_2CH_2Br \xrightarrow{NH_3} (CH_3)_2CHCH_2CH_2NH_2\cdot HBr$
   $\xrightarrow{NaOH} (CH_3)_2CHCH_2CH_2NH_2$

(3) $(CH_3)_2CHC\overset{H}{=}O \xrightarrow{HCN} (CH_3)_2CHCHCN \xrightarrow[heat]{H^{\oplus} -H_2O} (CH_3)_2CHCHCO_2H$
$\qquad\qquad\qquad\qquad\qquad\qquad\quad |\qquad\qquad\qquad\qquad\qquad\qquad |$
$\qquad\qquad\qquad\qquad\qquad\qquad\quad OH\qquad\qquad\qquad\qquad\qquad\qquad OH$

4. (1)
$$\begin{array}{cc}
\quad CH_3 & \quad CH_3 \\
H_2N-\overset{|}{C}-H & H-\overset{|}{C}-NH_2 \\
\quad \overset{|}{CO_2H} & \quad \overset{|}{CO_2H} \\
\text{D-alanine} & \text{L-alanine}
\end{array}$$

(2)
$$\begin{array}{cc}
\quad CH_3 & \quad CH_3 \\
\quad HCOH & \quad HOCH \\
H_2N\overset{|}{C}H & \quad HCNH_2 \\
\quad \overset{|}{CO_2H} & \quad \overset{|}{CO_2H} \\
\text{D-threonine} & \text{L-threonine}
\end{array}$$

$$\begin{array}{cc}
\quad CH_3 & \quad CH_3 \\
\quad HOCH & \quad HCOH \\
H_2N\overset{|}{C}H & \quad HCNH_2 \\
\quad \overset{|}{CO_2H} & \quad \overset{|}{CO_2H} \\
\text{D-allothreonine} & \text{L-allothreonine}
\end{array}$$

The formulas are also often drawn with the carboxyl at the top (rotate through 180°) to correspond to L-glyceride.

5. In order to obtain the zwitterion structure for an acidic amino acid, such as aspartic acid, where minimal migration in an electrical field would occur (isoelectric point), one carboxyl group must be undissociated. More acid conditions are necessary to attain this isoelectric point.

$$\begin{array}{ccc}
CO_2H & & CO_2^{\ominus} \\
| & & | \\
CH_2 & OH^{\ominus} & CH_2 \\
| & \underset{H^{\oplus}}{\rightleftharpoons} & | \\
H_3\overset{\oplus}{N}-\underset{H}{\overset{|}{C}}-CO_2^{\ominus} & & H_3\overset{\oplus}{N}-\underset{H}{\overset{|}{C}}-CO_2^{\ominus} \\
\text{acidic} & & \text{isoelectric point 2.87}
\end{array}$$

$$\begin{array}{ccc}
NH_2 & & \overset{\oplus}{NH_3} \\
| & & | \\
(CH_2)_4 & H^{\oplus} & (CH_2)_4 \\
| & \underset{OH^{\ominus}}{\rightleftharpoons} & | \\
H_3\overset{\oplus}{N}-\underset{H}{\overset{|}{C}}-CO_2^{\ominus} & & H_3\overset{\oplus}{N}-\underset{H}{\overset{|}{C}}-CO_2^{\ominus} \\
\text{basic} & & \text{isoelectric point 9.74}
\end{array}$$

For a basic amino acid such as lysine only one amino group is protonated in the zwitterion structure. Basic conditions are required to achieve this isoelectric point.

6.

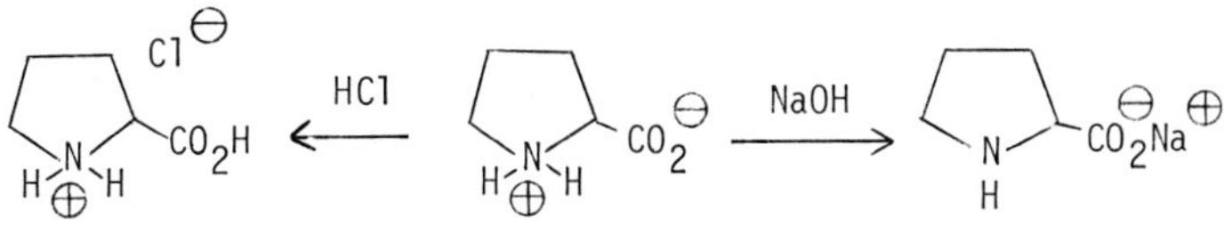

$$\underset{H \overset{\oplus}{\underset{H}{N}}}{\text{pyrrolidine}}-CO_2H \xleftarrow{HCl} \underset{H \overset{\oplus}{\underset{H}{N}}}{\text{pyrrolidine}}-CO_2^{\ominus} \xrightarrow{NaOH} \underset{\underset{H}{N}}{\text{pyrrolidine}}-CO_2^{\ominus}Na^{\oplus}$$

Fundamentally, these are two of the most important reactions of amino acids; compare with the equilibrium 17.12.

7.

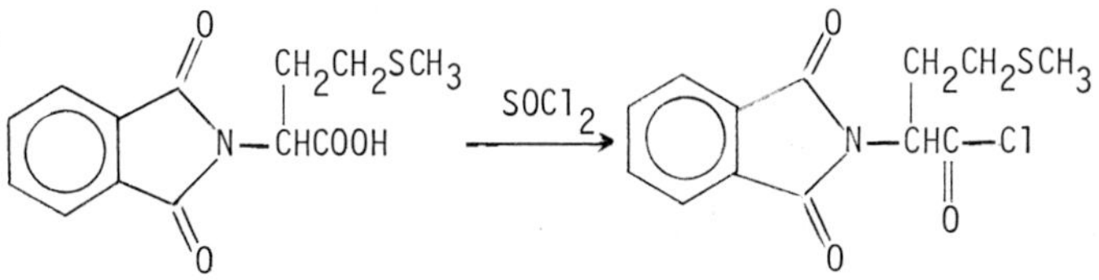

$$\text{phthalimide-N}-\underset{\underset{H}{|}}{\overset{CH_2CH_2SCH_3}{C}}HCOOH \xrightarrow{SOCl_2} \text{phthalimide-N}-\underset{\underset{O}{\parallel}}{\overset{CH_2CH_2SCH_3}{C}}HC-Cl$$

8. (1) 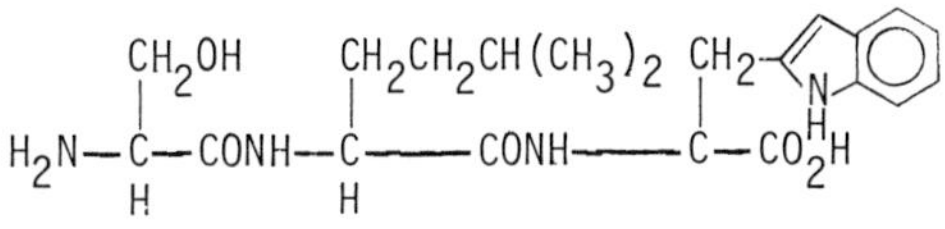 $H_2N-\underset{\underset{H}{|}}{\overset{CH_2OH}{C}}-CONH-\underset{\underset{H}{|}}{\overset{CH_2CH_2CH(CH_3)_2}{C}}-CONH-\underset{}{\overset{CH_2-\text{indole}}{C}}-CO_2H$

tripeptide, serylleucyltryptophan

(2) $H_2N-\underset{\underset{H}{|}}{\overset{CH_2-\text{imidazole}}{C}}-CONH-\underset{\underset{H}{|}}{\overset{CH_2CH(CH_3)_2}{C}}-CO_2H$

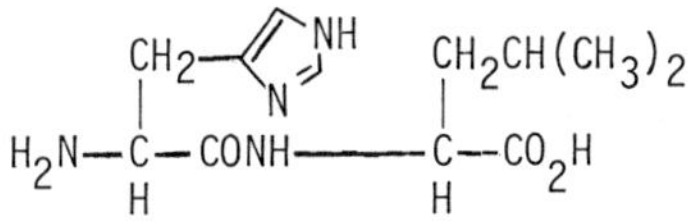

dipeptide, hystidylleucine

9.

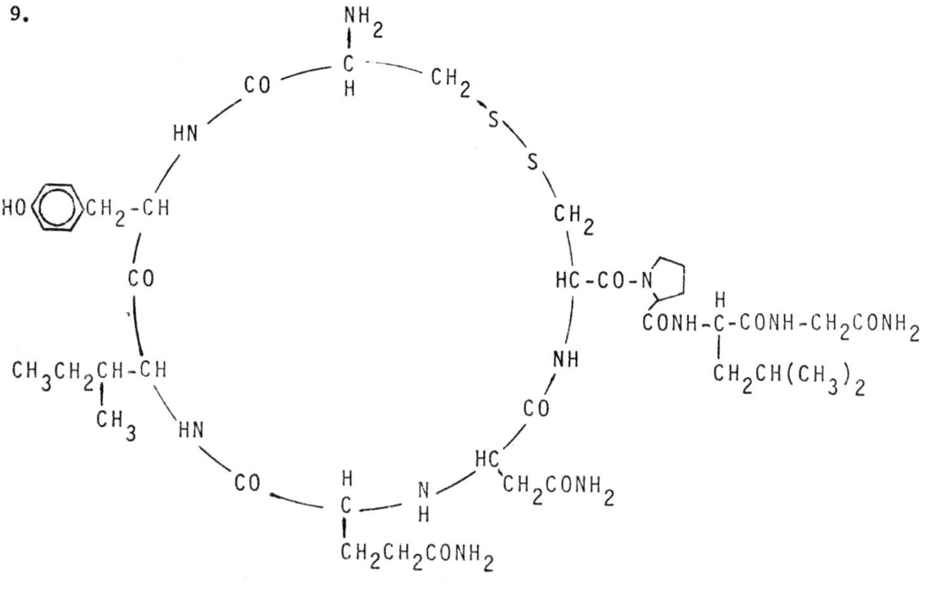

[cyclic peptide structure containing residues with side chains: CH_2-phenol-OH, $CH_3CH_2CH(CH_3)-$, $CH_2CH_2CONH_2$, CH_2CONH_2, pyrrolidine ring, $CH_2CH(CH_3)_2$, $CH_2-S-S-CH_2$, CH_2NH_2, with linking CO–NH amide bonds and terminal $CONH-CH_2CONH_2$]

10.

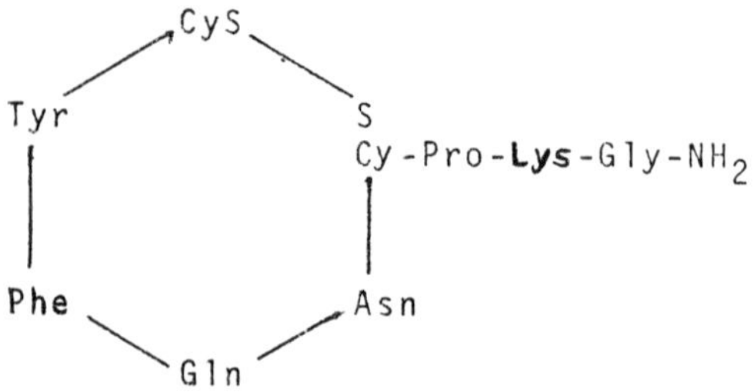

```
       Cys
      /   \
   Tyr     S
   |       |
           Cy-Pro-**Lys**-Gly-NH$_2$
   |       |
   Phe     Asn
      \   /
       Gln
```

11. Leu-(Gly)$_3$-Leu-(Gly)$_3$-Leu-(Gly)$_8$-Gly

12.

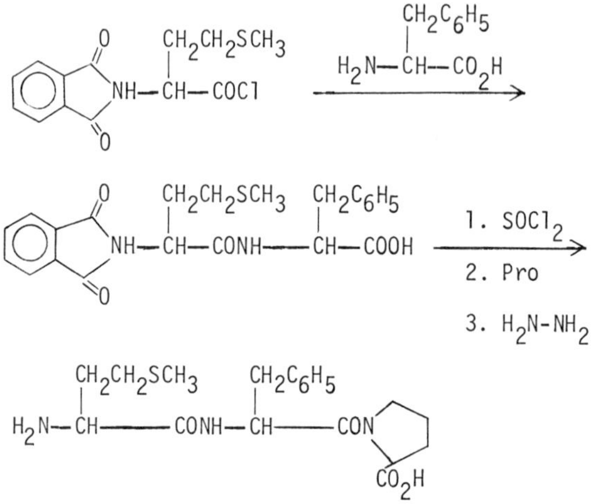

Phthalimide-NH—CH(CH$_2$CH$_2$SCH$_3$)—COCl + H$_2$N—CH(CH$_2$C$_6$H$_5$)—CO$_2$H →

Phthalimide-NH—CH(CH$_2$CH$_2$SCH$_3$)—CONH—CH(CH$_2$C$_6$H$_5$)—COOH

1. SOCl$_2$
2. Pro
3. H$_2$N-NH$_2$

H$_2$N—CH(CH$_2$CH$_2$SCH$_3$)—CONH—CH(CH$_2$C$_6$H$_5$)—CON(Pro)—CO$_2$H

13.

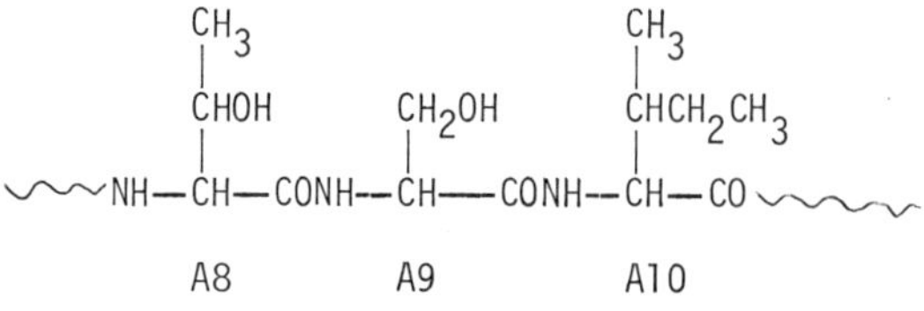

~~~NH—CH(CHOHCH$_3$)—CONH—CH(CH$_2$OH)—CONH—CH(CHCH$_2$CH$_3$ with CH$_3$)—CO~~~
        A8   A9   A10

human insulin

~~~NH—CH(CH$_3$)—CONH—CH$_2$—CONH—CH(CH(CH$_3$)$_2$)—CO~~~
    A8   A9   A10

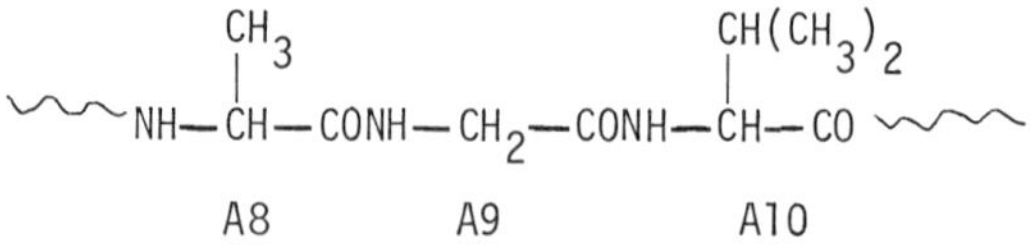

sheep insulin

14. A chain

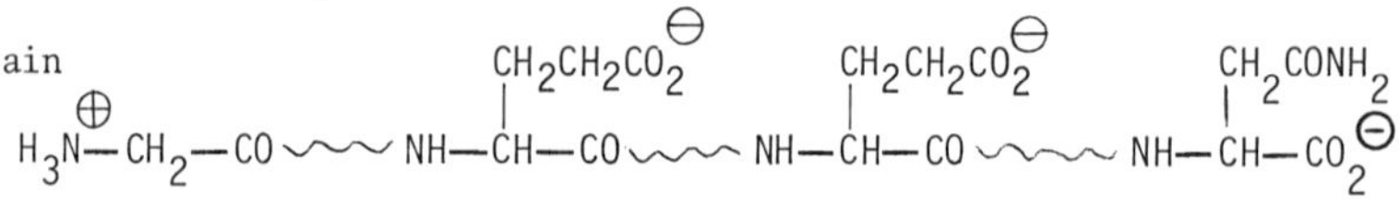

H$_3$N$^{\oplus}$—CH$_2$—CO~~~NH—CH(CH$_2$CH$_2$CO$_2^{\ominus}$)—CO~~~NH—CH(CH$_2$CH$_2$CO$_2^{\ominus}$)—CO~~~NH—CH(CH$_2$CONH$_2$)—CO$_2^{\ominus}$

B chain

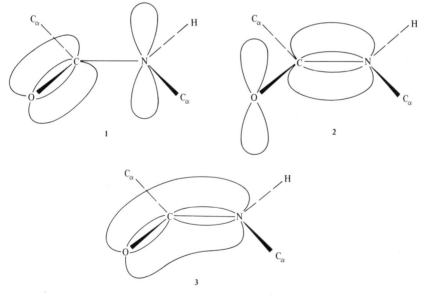

15. The molecular orbital representations (1) and (2) correspond to the structures in Formula 17.23. The actual structure involves π orbitals overlapping all three atoms leading to restricted rotation of the C-N bond or a planar configuration of the peptide linkage as shown in (3) (Fig. A-17.15).

Figure A-17.15

16. Oxidation of the thiol groups to disulfide linkages results in a natural folding of the peptide chain.

17.

$$\underset{\text{phenylpyruvic acid}}{\underset{O=C-CO_2H}{\overset{CH_2-C_6H_5}{|}}} \xleftarrow{\text{enzyme deamination}} \underset{}{\underset{H_2N-CH-CO_2H}{\overset{CH_2-C_6H_5}{|}}} \xcancel{\xrightarrow{\text{enzyme}}} \underset{}{\underset{H_2N-CH-CO_2H}{\overset{CH_2-C_6H_4-OH}{|}}}$$

Apparently the overproduction of phenylpyruvic acid is a result of an excessive amount of phenylalanine being deaminated. This in turn results from a blockage of the conversion of phenylalanine to tyrosine by the inavailability of the enzyme which promotes this oxidation of the benzene ring. This is a genetic disease or a so-called "inborn error in metabolism." The serious result is feeble-mindedness in an afflicted infant; if recognized early, it can be controlled by limiting phenylalanine intake.

18. phenylalanine: UUU, UUC   valine: GUU, GUC, GUA, GUG

19.

(1) $H_2N-\underset{H}{\overset{CH_2C_6H_5}{\underset{|}{C}}}-\underset{\|}{\overset{}{C}}-NH-\underset{H}{\overset{CH_2CH(CH_3)_2}{\underset{|}{C}}}-\underset{\|}{\overset{}{C}}-NH-\underset{H}{\overset{CH_2CONH_2}{\underset{|}{C}}}-CO_2H$
     $\qquad\qquad\qquad\qquad O \qquad\qquad\qquad\qquad O$

Phe-Leu-Asn, a tripeptide

(2) $H_2N-\underset{H}{\overset{\overset{CH_3}{|}\;CHCH_2CH_3}{\underset{|}{C}}}-\underset{\|}{\overset{}{C}}-NH-\underset{H}{\overset{(CH_2)_4NH_2}{\underset{|}{C}}}-\underset{\|}{\overset{}{C}}-NH-\underset{H}{\overset{(CH_2)_2CO_2H}{\underset{|}{C}}}-\underset{\|}{\overset{}{C}}-NH-\underset{H}{\overset{CH_2CH_2SCH_3}{\underset{|}{C}}}-CO_2H$

Ile-Lys-Glu-Met, a tetrapeptide

(3) $H_2N-\underset{H}{\overset{CH_2-C_6H_4-OH}{\underset{|}{C}}}-\underset{\|}{\overset{}{C}}-NH-CH_2-\underset{\|}{\overset{}{C}}-NH-CH_2-\underset{\|}{\overset{}{C}}-NH-\underset{H}{\overset{CH_2-\text{imidazole}}{\underset{|}{C}}}-\underset{\|}{\overset{}{C}}-N\text{(pyrrolidine)}-CO_2H$

Tyr-Gly-Gly-His-Pro, a pentapeptide

(4) $H_2N-\underset{H}{\overset{\overset{CH_3}{|}\;CHOH}{\underset{|}{C}}}-\underset{\|}{\overset{}{C}}-NH-\underset{H}{\overset{CH_2CONH_2}{\underset{|}{C}}}-\underset{\|}{\overset{}{C}}-NH-\underset{H}{\overset{(CH_2)_2CO_2H}{\underset{|}{C}}}-CO_2H$

Thr-Asn-Glu, a tripeptide

20. (1) The pH at which there is a minimal tendency for an amino acid species in an electric field to migrate toward either electrode.

(2) An N-terminal residue is the amino acid unit at the end of the polypeptide chain, which has a free α-amino group.

$$H_2N-\underset{H}{\overset{CH_3}{\underset{|}{C}}}-\underset{\|}{\overset{}{C}}\sim\sim\sim$$

(3) Globular proteins are soluble proteins; they range widely in molecular weight—from those examples in Table 17.1 from insulin (11,600) to glutamic dehydrogenase (2,000,000).

(4) A polypeptide is an amino acid polyamide containing between approximately 10 and 100 residues.

$$H_2N-\underset{H}{\overset{R}{\underset{|}{C}}}-\underset{\|}{\overset{}{C}}-\left[NH-\underset{H}{\overset{R}{\underset{|}{C}}}-\underset{\|}{\overset{}{C}}\right]_n-NH-\underset{H}{\overset{R}{\underset{|}{C}}}-CO_2H$$

The R side chains are representative of any of the amino acids randomly distributed.

(5) Essential amino acids are those that cannot be synthesized by the host and thus must be present in adequate quantity in the protein of the diet. Those essential for man are: Val, Leu, Ile, Phe, Thr, Met, Lys, Trp.

(6) Secondary protein structure is the conformation which a polypeptide chain assumes as a result of hydrogen bonding between carbonyl oxygens and amino hydrogens of the peptide linkages connecting the amino acid residues and the coplanar nature of the amide group. A common secondary structure is the α-helix.

(7) Tertiary protein structure results from the specific manner in which the polypeptide chain is folded. This results from hydrophobic forces involving the side chains and disulfide bridging of cysteine residues to stabilize the structure once it is formed.

(8) Denaturation involves the breakdown of protein structure, principally quaternary through secondary, as a result of heat, electromagnetic irradiations, the action of chemicals, and the like. The process is often irreversible as in the case of cooking eggs or meat.

(9) Fibrous proteins are insoluble in water, as would be expected for proteins that serve a function of structural material in animal life, just as cellulose does for plants. Examples are keratin in skin, hair, wool, horn, and feathers; collagen in tendons; and myosin in muscle.

(10) Conjugated proteins consist of protein and nonprotein moieties bound together. Examples are nucleoprotein (protein·nucleic acid), glycoprotein (protein·carbohydrate), lipoprotein (protein·lipid), and chromoprotein (protein·metal).

21.

$$H_2N-\overset{H}{\underset{}{C}}(\overset{CH_3}{})-\overset{O}{\underset{}{C}}-\overset{H}{\underset{}{N}}-\overset{H}{\underset{}{C}}(\overset{CO_2H}{CH(CH_3)_2})$$

(circles around the two α-carbons)

22.

(1) $(CH_3)_2CHCH_2CH_2COOH \xrightarrow[\text{2. } H_2O]{\text{1. } Br_2\text{-P}} (CH_3)_2CHCH_2\underset{Br}{CHCO_2H}$

$\xrightarrow{NH_3} (CH_3)_2CHCH_2\underset{NH_2}{CHCO_2H}$

(2) $HC\equiv CH \xrightarrow[H^\oplus - H_2O]{Hg^{2\oplus}} CH_3C=O \xrightarrow[NaCN]{NH_4Cl^-} CH_3\underset{NH_2}{\overset{H}{C}}-CN$

$\xrightarrow[\text{heat}]{H^\oplus -H_2O} CH_3\underset{NH_2}{CHCO_2H}$

(3) $\text{Ph}-CH_3 \xrightarrow[\text{uv light}]{Cl_2} \text{Ph}-CH_2Cl \xrightarrow[\text{dry } Et_2O]{Mg} \text{Ph}-CH_2MgCl$

$\xrightarrow[\text{2. } H^\oplus -H_2O]{\text{1. } HC=O/H} \text{Ph}-CH_2CH_2OH \xrightarrow[\text{heat}]{Cu} \text{Ph}-CH_2-C=O/H$

$\xrightarrow[NaCN]{NH_4Cl} \text{Ph}-CH_2\underset{NH_2}{CH}-CN \xrightarrow{H^\oplus -H_2O} \text{Ph}-CH_2\underset{NH_2}{CHCO_2H}$

(4) $(CH_3)_2CHCH_2OH \xrightarrow[\text{heat}]{Cu} (CH_3)_2CHC=O/H \xrightarrow[NaCN]{NH_4Cl}$

$(CH_3)_2CHCHCN \atop NH_2 \xrightarrow[\text{heat}]{H^\oplus -H_2O} (CH_3)_2CH\underset{NH_2}{CHCO_2H}$

23. Arg-Pro-Pro-Gly-Phe-Ser-Pro-Phe-Arg

Bradykinin

24.

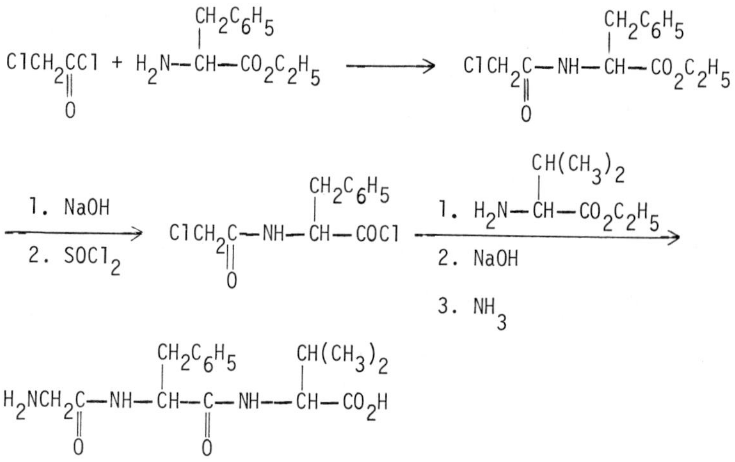

$$ClCH_2CCl + H_2N-CH(CH_2C_6H_5)-CO_2C_2H_5 \longrightarrow ClCH_2C(O)-NH-CH(CH_2C_6H_5)-CO_2C_2H_5$$

$$\xrightarrow[\text{2. SOCl}_2]{\text{1. NaOH}} ClCH_2C(O)-NH-CH(CH_2C_6H_5)-COCl \xrightarrow[\substack{\text{2. NaOH} \\ \text{3. NH}_3}]{\text{1. H}_2\text{N-CH(CH(CH}_3)_2)\text{-CO}_2\text{C}_2\text{H}_5}$$

$$H_2NCH_2C(O)-NH-CH(CH_2C_6H_5)-C(O)-NH-CH(CH(CH_3)_2)-CO_2H$$

25.

(1) 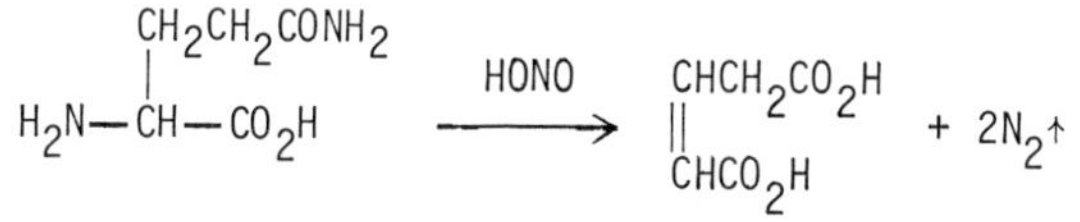 $H_2N-CH(CH_2CH_2CONH_2)-CO_2H \xrightarrow{HONO} CHCH_2CO_2H \text{ (=CHCO}_2\text{H)} + 2N_2\uparrow$

(2) $H_2N(CH_2)_4CH(NH_2)CO_2H \xrightarrow[\text{NaOH}]{C_6H_5COCl} C_6H_5CONH(CH_2)_4CH(NHCOC_6H_5)CO_2H$

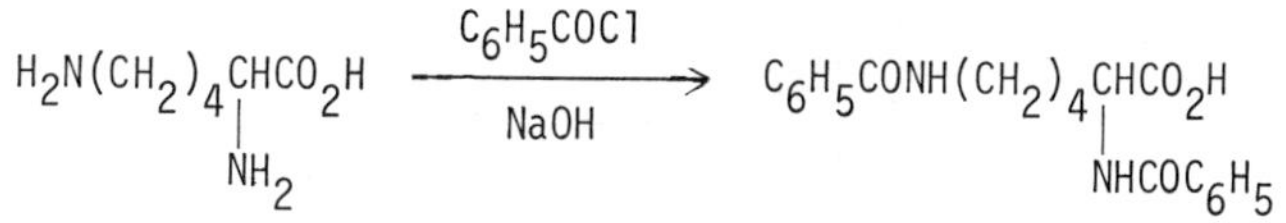

(3) 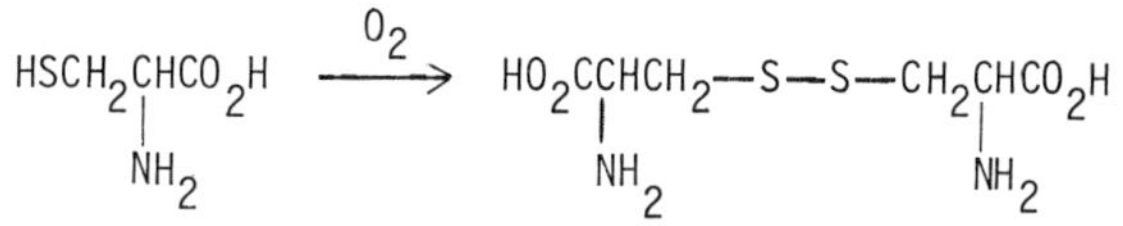 $HSCH_2CH(NH_2)CO_2H \xrightarrow{O_2} HO_2CCH(NH_2)CH_2-S-S-CH_2CH(NH_2)CO_2H$ cystine Cys-Cys

(4) 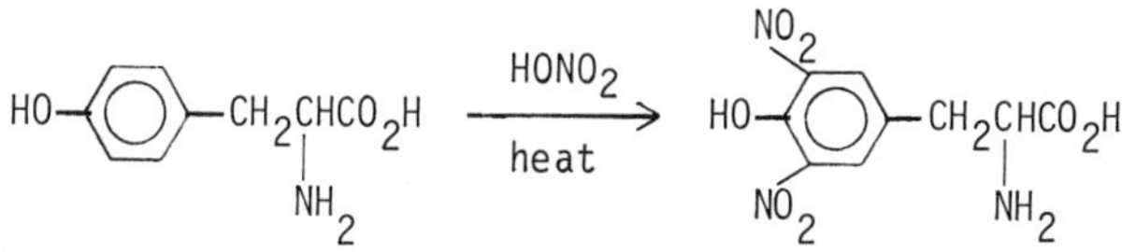 $HO-\langle\bigcirc\rangle-CH_2CH(NH_2)CO_2H \xrightarrow[\text{heat}]{HONO_2} HO-\langle\bigcirc(NO_2)_2\rangle-CH_2CH(NH_2)CO_2H$

(5) 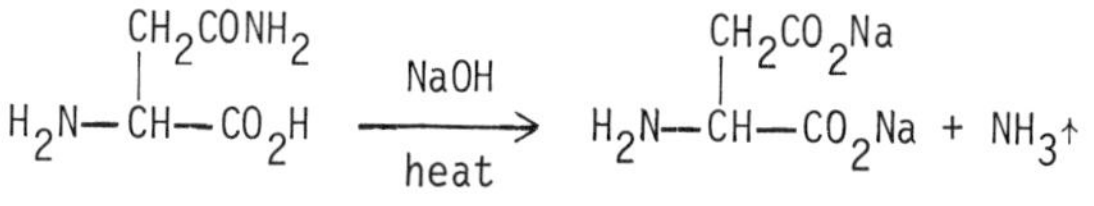 $H_2N-CH(CH_2CONH_2)-CO_2H \xrightarrow[\text{heat}]{NaOH} H_2N-CH(CH_2CO_2Na)-CO_2Na + NH_3\uparrow$

(6) 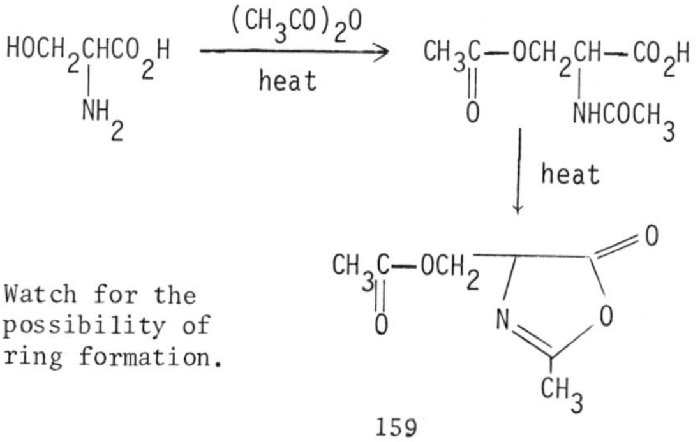 $HOCH_2CH(NH_2)CO_2H \xrightarrow[\text{heat}]{(CH_3CO)_2O} CH_3C(O)-OCH_2CH(NHCOCH_3)-CO_2H$

$\downarrow$ heat

[oxazoline ring: $CH_3C(O)-OCH_2-$ attached to ring with N, O, CH$_3$, =O]

Watch for the possibility of ring formation.

26. Ser-Asp-Asn-Asn-Gln-Gln-Gly-Lys-Ser-Ala-Gln-Gln-Gly-Gly-Tyr

    scotophobin

27. $(CH_3)_2\underset{NH_2}{C}-CO_2H$, 2-amino-2-methylpropionic acid

    $CH_3\underset{NHCH_3}{CH}CO_2H$, 2-(N-methylamino)propionic acid

    $C_2H_5NHCH_2CO_2H$, N-ethylaminoacetic acid

    $H_2NCH_2CH_2CO_2H$, 3-aminopropionic acid

    $CH_3\underset{NH_2}{CH}CH_2CO_2H$, 3-aminobutyric acid

28. (1)

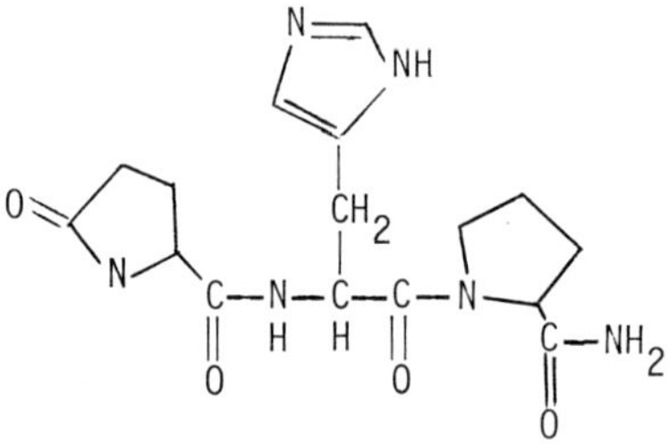

    (2)

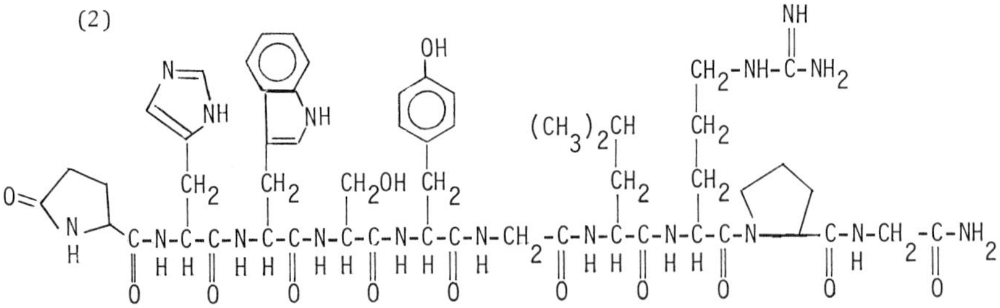

chapter 18

NATURAL PRODUCTS AND PHYSIOLOGICALLY ACTIVE SUBSTANCES

1.

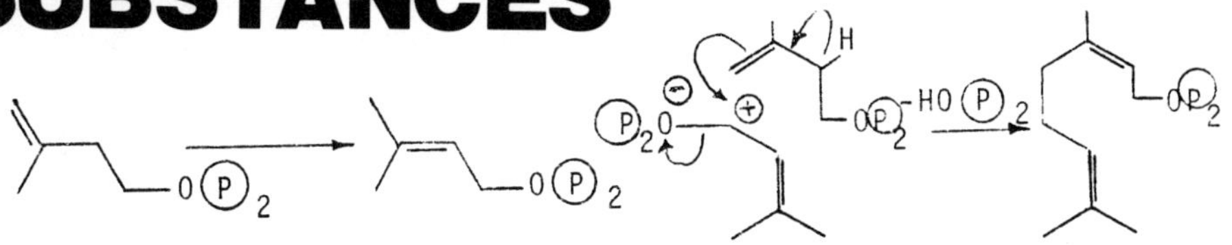

2.  α-pinene

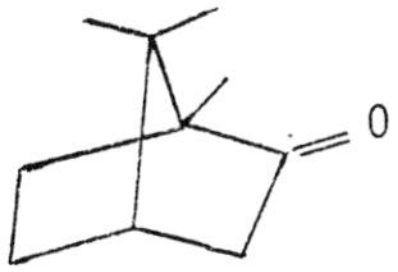

   camphor

3. 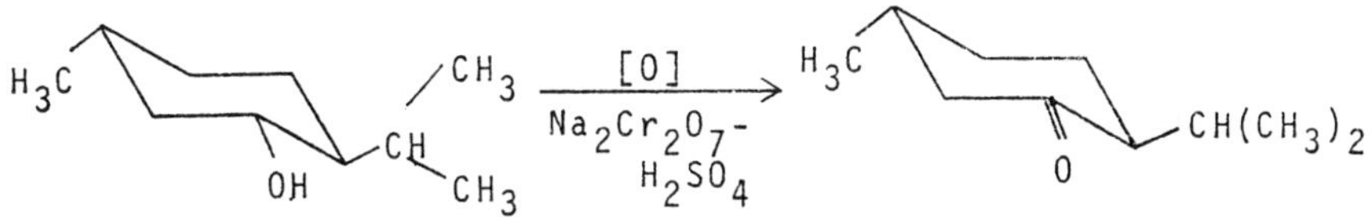 menthol → [O], $Na_2Cr_2O_7$-H_2SO_4 → menthone

4. 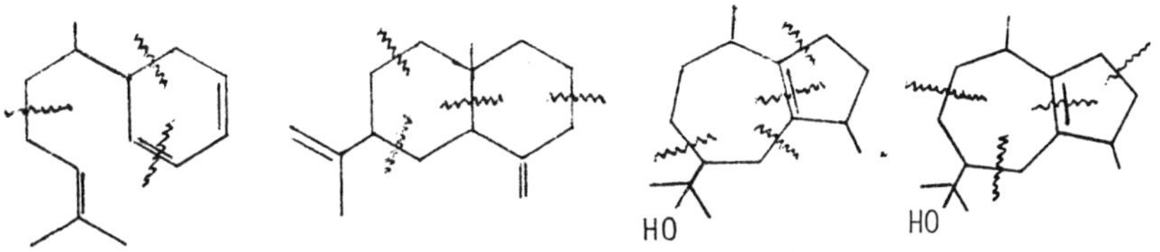 Either is correct.

5.

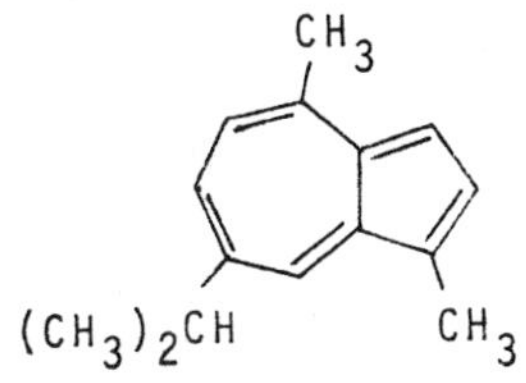

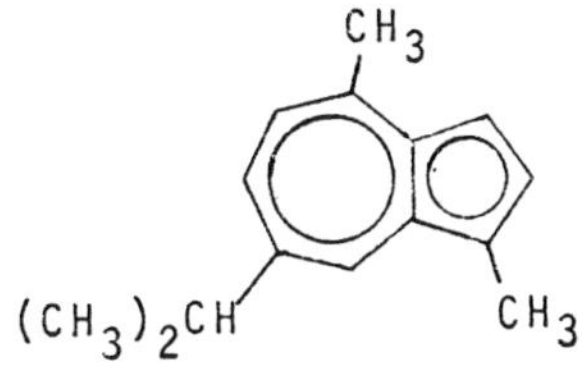

Note that the azulene ring system can be represented by alternating double and single bonds or by the circles in each ring. Note also that where in benzene n = 1 for 4n + 2, that is, 6π electrons, azulene is a case where n = 2 (10π electrons) and thus would be predicted to be aromatic.

6.

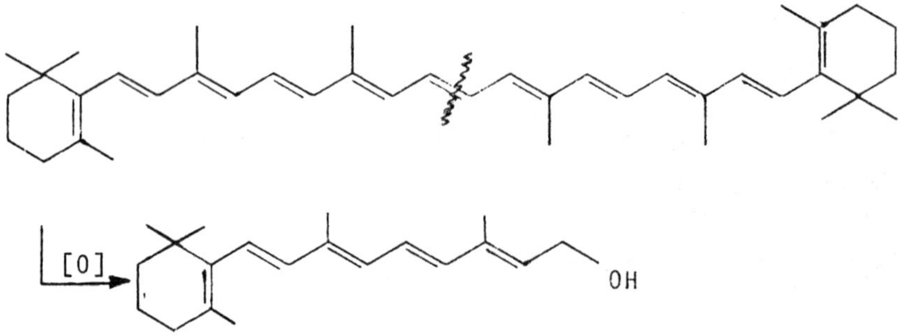

[O]
OH

7.

hν
HO
CH2
HO
vitamin D3

8.

CH3
HO
CH3
OH
CH3
-HO(P)2
electrophilic substitution of phytol pyrophosphate
CH3
HO
CH3
CH3
H
nucleophilic addition of hydroxyl group to double bond according to Markownikoff's rule

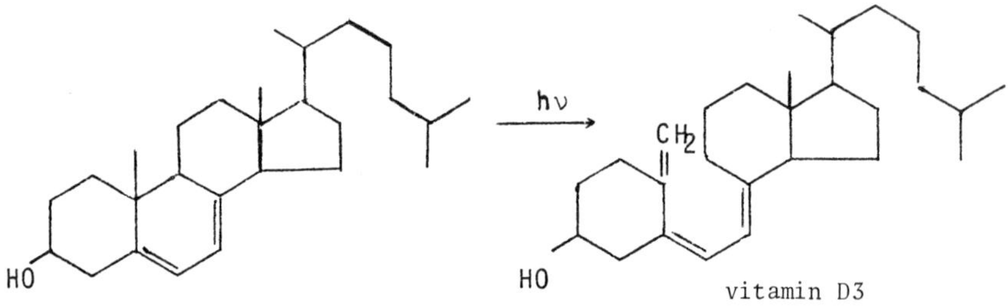

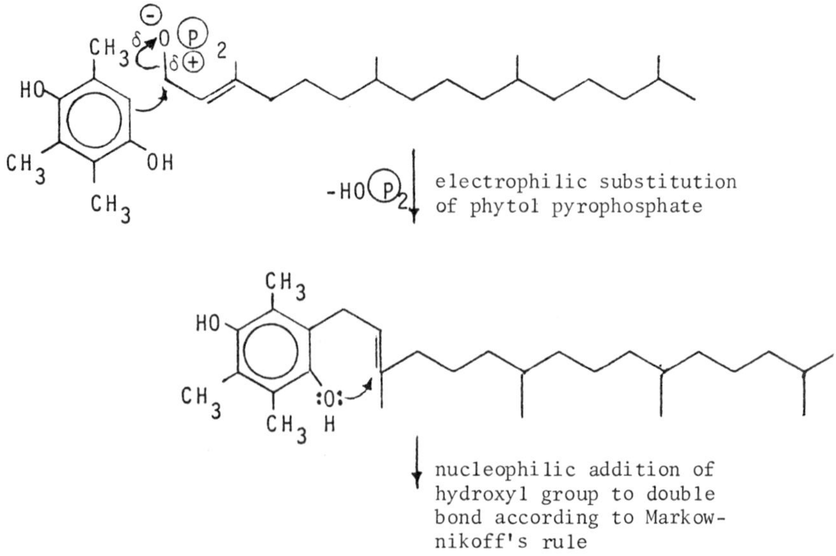

9.

(1) [structure: two chromone/coumarin-like rings connected by -CH₂- with OH groups]

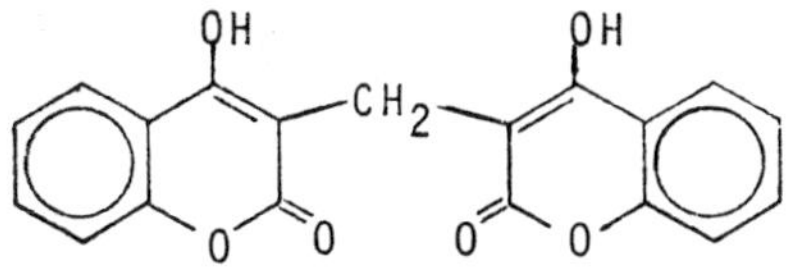

(2) [structure: chromone with OH and -CHCH₂COCH₃ side chain bearing C₆H₅]

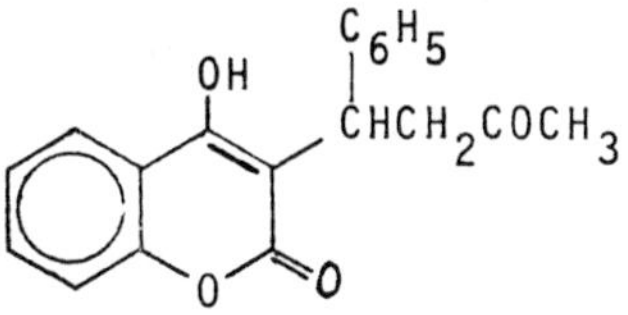

10.

[structure on left with H₃C groups, N's, NH, =O] ⇌ (2[H] / [O]) [structure on right with H₃C groups, NH, N, =O]

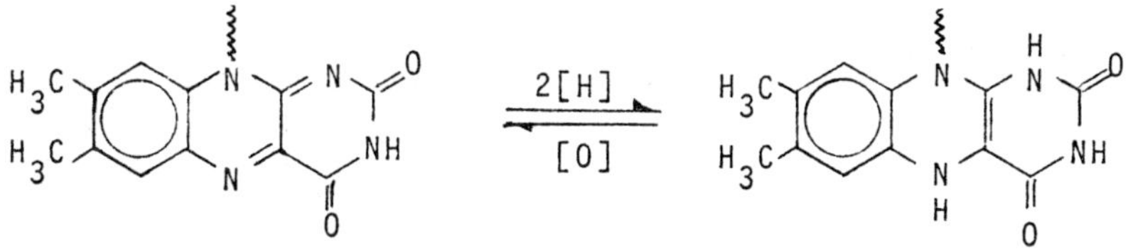

11. (1) Draw out four pyrrole rings in the positions indicated and remember that they are partially saturated.

(2) Connect the pyrrole rings with a methine bridge except between rings A and D; the rings are connected through position 2 of each ring. Substitute a methyl group at the top and bottom methine bridges.

(3) Position the cobalt atom in the center and draw one conventional bond from the N of ring A and three coordinate bonds from the N's of rings B, C, and D to the cobalt. Finally, attach a cyano group to the cobalt.

(4) Starting with pyrrole ring A, attach a propionamide substituent through position 3 of the chain to position 3 of the four rings. Ring D only also has a CH₃ substituent at position 3.

(5) Return now to pyrrole ring A. In position 5, place a methyl substituent; this is the only 5- substituent in the molecule. Attach an acetamide group through position 2 of its chain to position 4 of pyrrole rings A and B. Also attach a CH₃ substituent at position 4 in rings A and B. Attach two CH₃ substituents at position 4 in ring C. Then in ring D introduce only an acetamide substituent in position 4.

(6) Starting at the top methine bridge, introduce a double bond on the left-hand side, and then continue with conjugated double bonds clockwise to ring D. This is one of two resonance structures. The other results by shifting the double bond in ring D one position to the right onto the methine bridge and continuing counterclockwise into ring A.

(7) On the propionamide substituent in ring D, consider that the amide linkage has been formed through the amino group of 1-amino-2-propanol. A ribonucleotide function is then attached through the 2-hydroxyl group of the 1-amino-2-propanol residue. The ribose element of structure is then attached through an

α-glycosidic linkage to the 5,6-dimethylbenzimidazole group, which in turn then coordinates with the cobalt atom through the other benzimidazole nitrogen atom.

12.

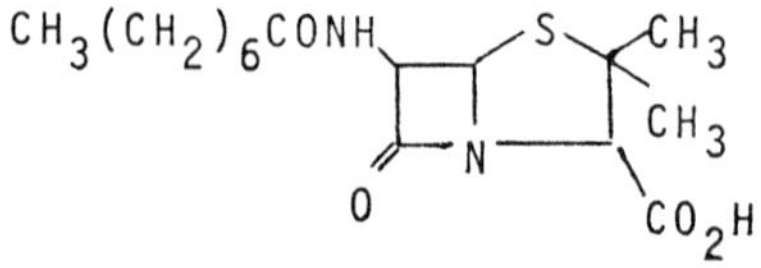

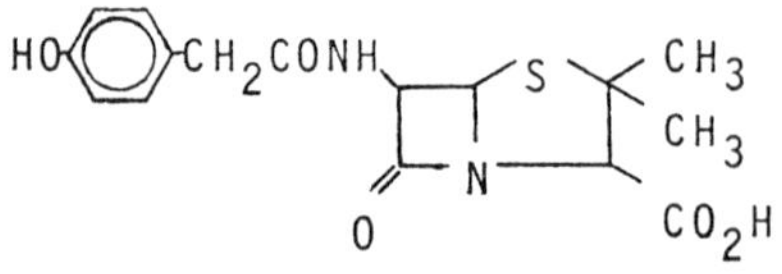

13.

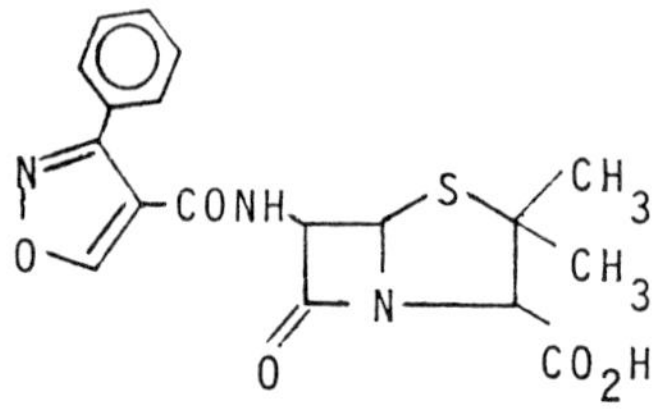

oxacillin

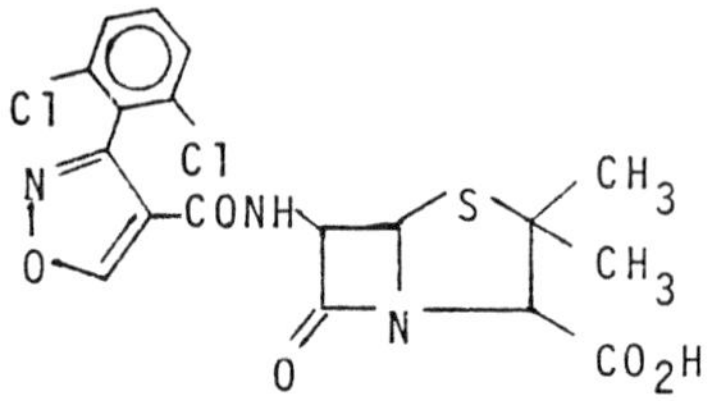

dicloxacillin

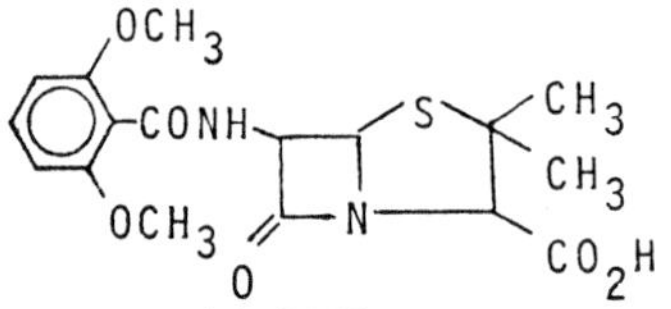

methicillin

14. The only group which differs from those found in benzylpenicillin is the amido group.

15.

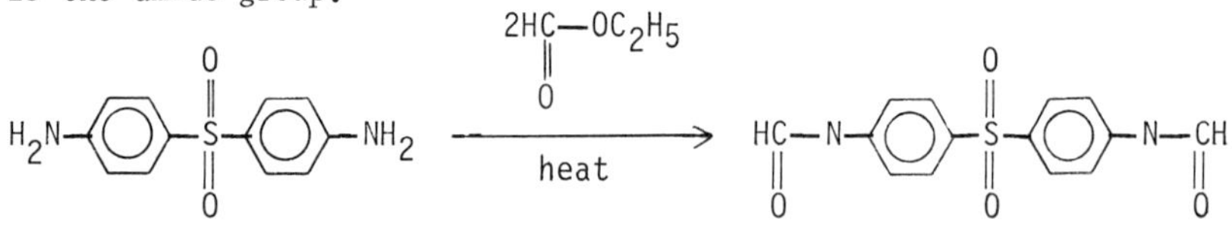

DDS

An ester must be used for this acylation because the acid chloride and anhydride of formic acid are unstable.

16.

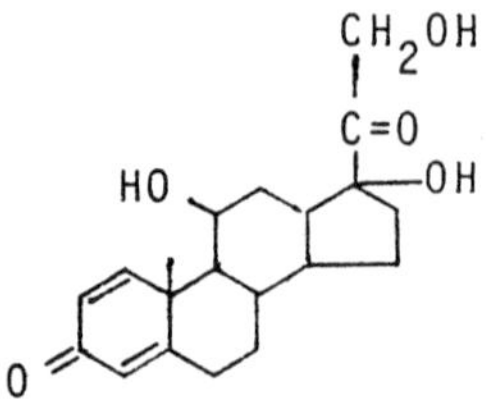

prednisolone

17.

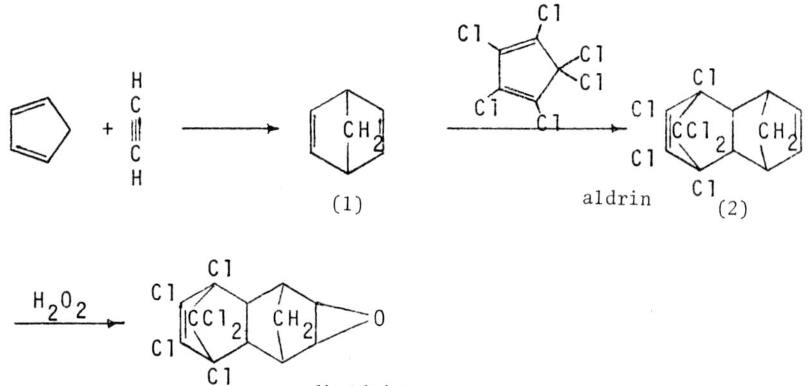

[Structures: cyclopentadiene + CH₂=CH—CH=CH₂ → (1) bicyclic structure with CH₂; then + hexachlorocyclopentadiene → aldrin (2); then H₂O₂ → (3) dieldrin]

18.

$(C_2H_5O)_2P-O$—[coumarin structure with CH_3 and Cl, C=O, O in ring]
        ‖
        S

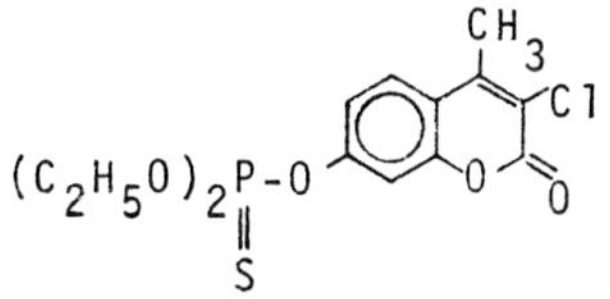

19. (1) gypsy moth pheromone: (Z)-7-hexadecene-1,10-diol 10-acetate (CA).
(2) gyplure: 12-acetoxy-cis-9-octadecen-1-ol.   (3) oriental fruit fly: 4-allyl-1,2-methoxybenzene, methyleugenol.   (4) Mediterranean fruit fly: (e)-tert-butyl-5(e)-chloro-trans-2(e)-methyl-cyclohexanecarboxylate.

20.

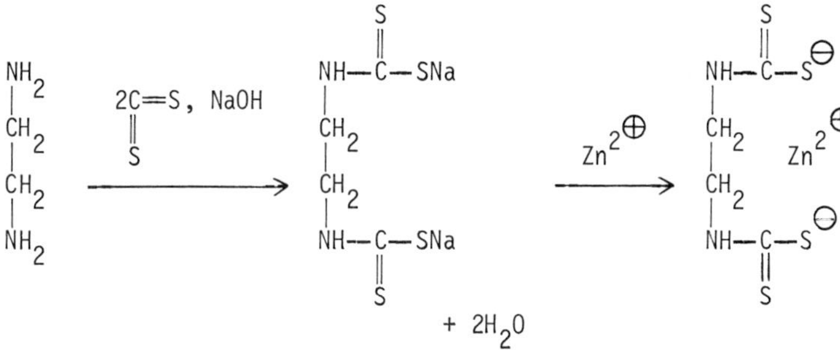

$$\begin{matrix} NH_2 \\ | \\ CH_2 \\ | \\ CH_2 \\ | \\ NH_2 \end{matrix} \xrightarrow{2C=S,\ NaOH} \begin{matrix} NH-\overset{S}{\underset{\|}{C}}-SNa \\ | \\ CH_2 \\ | \\ CH_2 \\ | \\ NH-\underset{\|}{\overset{S}{C}}-SNa \end{matrix} + 2H_2O \xrightarrow{Zn^{2+}} \begin{matrix} NH-\overset{S}{\underset{\|}{C}}-S^{\ominus} \\ | \\ CH_2 \\ | \\ CH_2 \\ | \\ NH-\underset{\|}{\overset{S}{C}}-S^{\ominus} \end{matrix} Zn^{2+}$$

Nabam

Must be made through the sodium salt because zinc hydroxide is insoluble.

21.

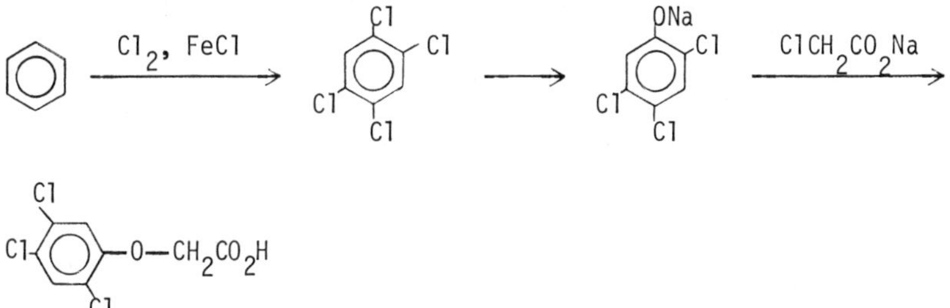

[Benzene] $\xrightarrow{Cl_2,\ FeCl}$ [1,2,4-trichlorobenzene] → [2,4,5-trichlorophenol sodium salt (ONa)] $\xrightarrow{ClCH_2CO_2Na}$

[2,4,5-trichlorophenoxyacetic acid: Cl₃C₆H₂—O—CH₂CO₂H]

22.

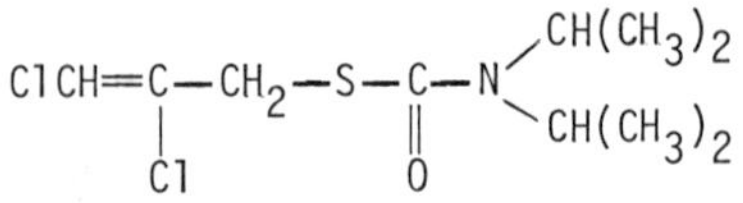

$$ClCH=C(Cl)-CH_2-S-C(=O)-N(CH(CH_3)_2)_2$$

Avadex

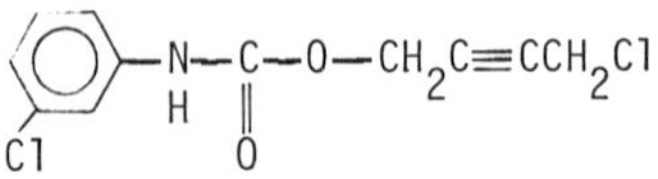

$$\text{(Cl-C}_6\text{H}_4\text{)}-N(H)-C(=O)-O-CH_2C\equiv CCH_2Cl$$

Barbam

23.

(1) 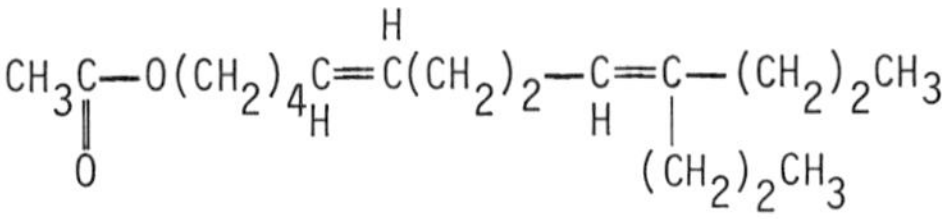 $CH_3C(=O)-O(CH_2)_4\underset{H}{C}=\underset{H}{C}(CH_2)_2-\underset{H}{C}=\underset{|}{C}-(CH_2)_2CH_3$
    with $(CH_2)_2CH_3$ substituent

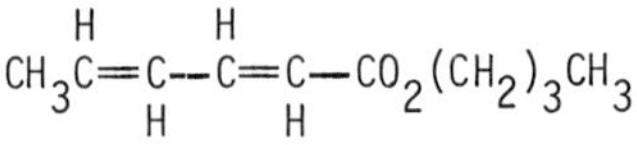

(2) $CH_3\underset{}{\overset{H}{C}}=\underset{H}{C}-\underset{H}{C}=\underset{H}{C}-CO_2(CH_2)_3CH_3$

(3) $CH_3C(=O)-O(CH_2)_6\underset{H}{C}=\underset{H}{C}(CH_2)_3CH_3$

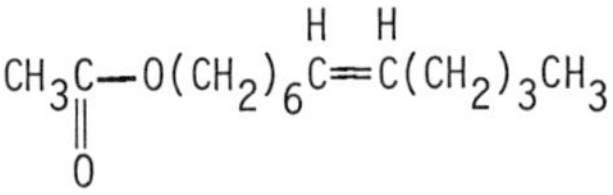

(4) 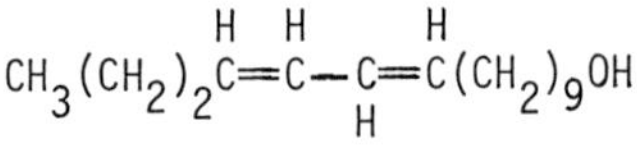 $CH_3(CH_2)_2\underset{H}{C}=\underset{H}{C}-\underset{H}{C}=C(CH_2)_9OH$

(5) 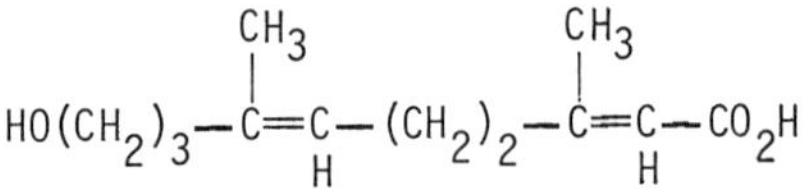 $HO(CH_2)_3-\underset{H}{\overset{CH_3}{C}}=C-(CH_2)_2-\underset{H}{\overset{CH_3}{C}}=C-CO_2H$

(6) 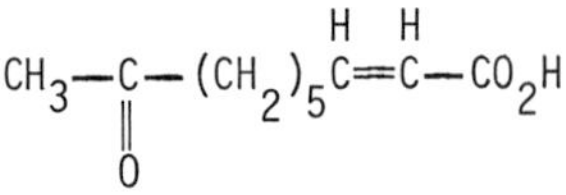 $CH_3-C(=O)-(CH_2)_5\underset{H}{C}=\underset{H}{C}-CO_2H$

(7) 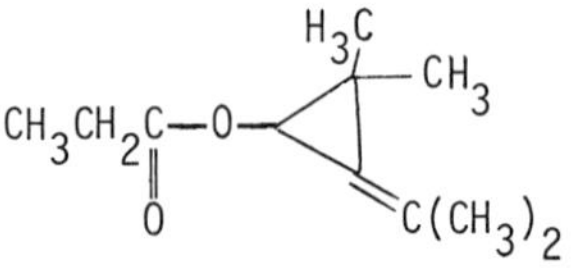 $CH_3CH_2C(=O)-O-$ [cyclopropane ring with H_3C, CH_3, and $C(CH_3)_2$ substituents]

24. Numbers in the starting materials refer to their eventual position in the heterocyclic ring and do not refer to the naming of the starting materials.

$$\underset{1\,NH_2}{CH_3-\overset{2}{C}=\overset{3}{N}H} \;+\; \begin{array}{c}\overset{4}{COOC_2H_5}\\ \overset{5}{C}HCH_2OC_2H_5\\ \overset{6}{H}C=O\end{array} \xrightarrow{NaOC_2H_5} \text{[pyrimidine: 1-N, 2-CH}_3\text{, 3-N, 4-OH, 5-CH}_2OC_2H_5\text{, 6-H]}$$

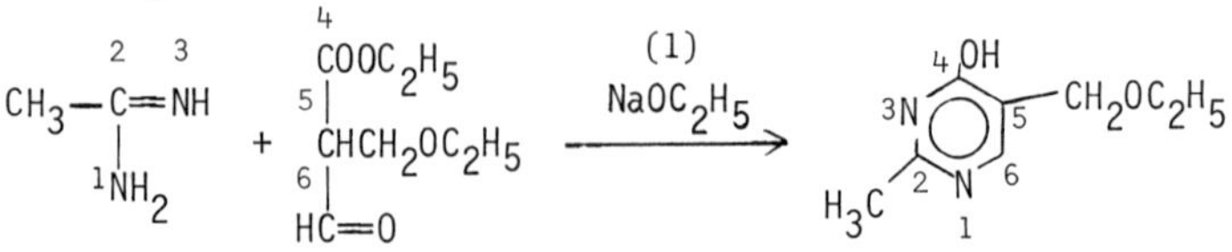

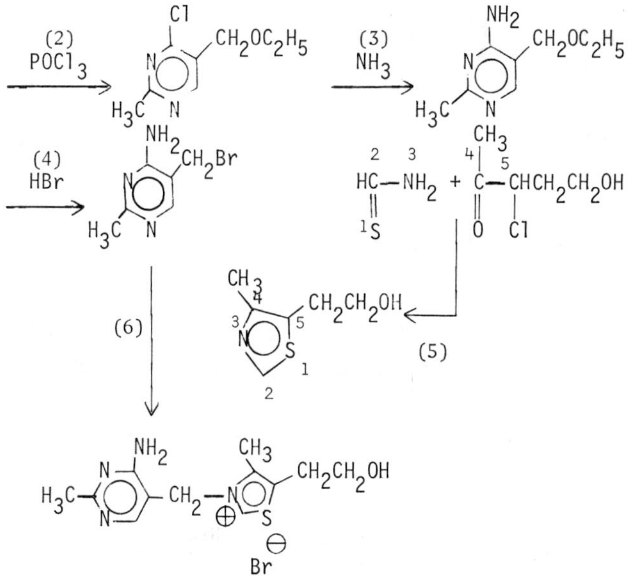

25.

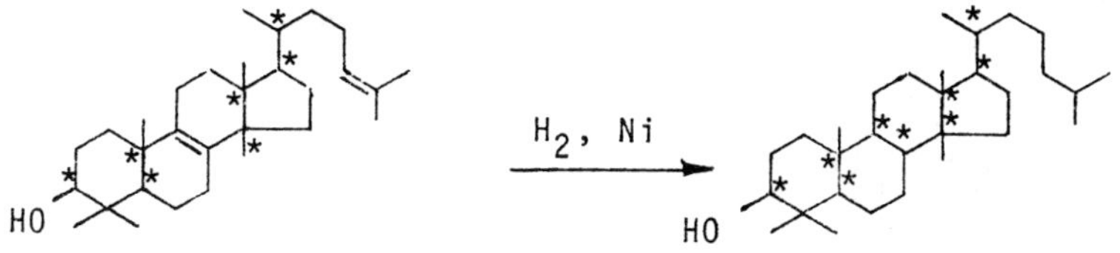

(1) $2^7 = 128$ (2) $2^9 = 512$

(3) The double bond in the side chain does not result in geometric isomerism because of the like terminal methyl groups. Nor does the double bond in the ring pose such possibilities because only rings C_8 or larger are capable of existing in <u>trans</u> forms.

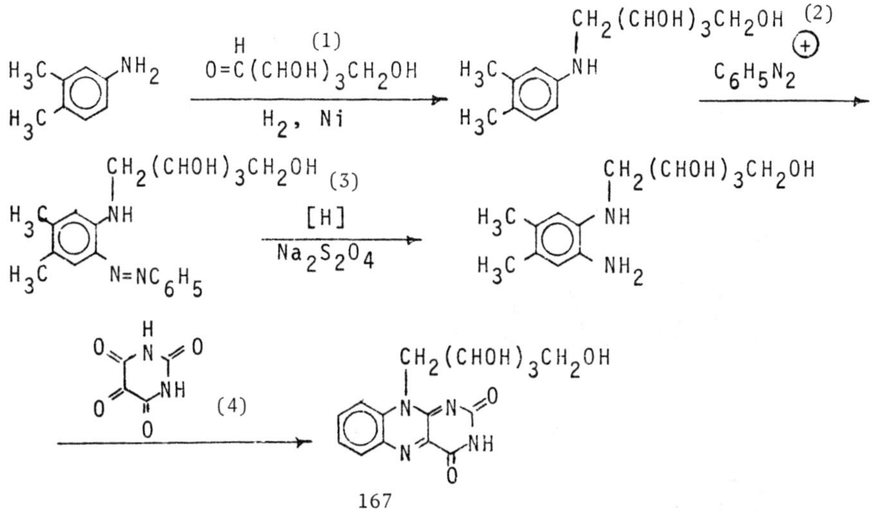

27.

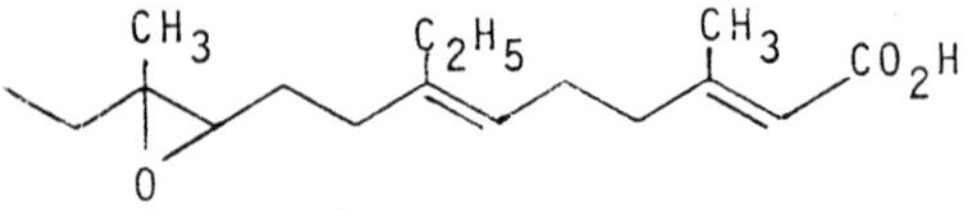

[Structure of juvenile hormone]

juvenile hormone

28.

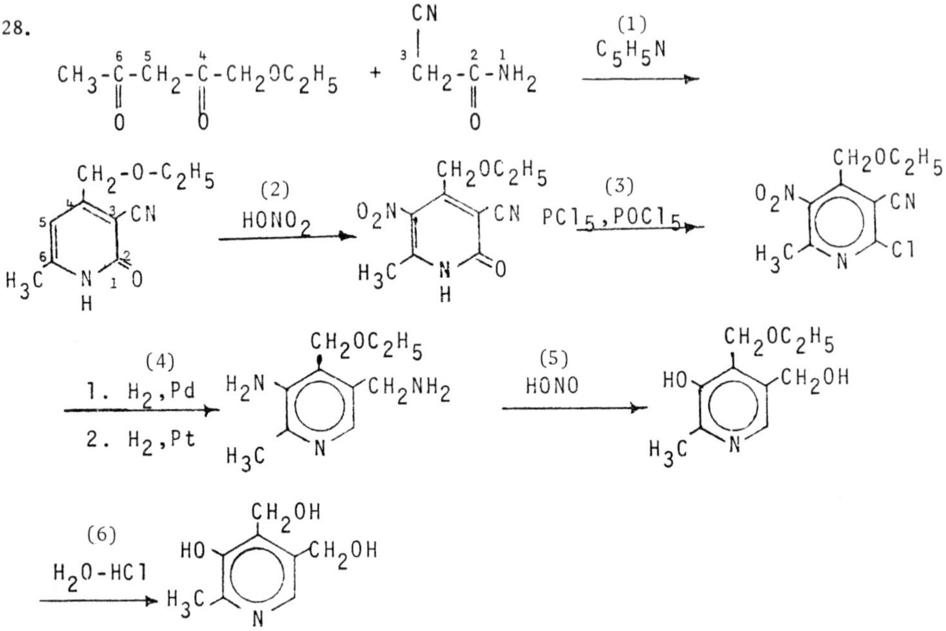

$CH_3-\overset{6}{\underset{O}{C}}-\overset{5}{C}H_2-\overset{4}{\underset{O}{C}}-CH_2OC_2H_5$ + $\overset{3}{C}H_2-\overset{2}{\underset{O}{C}}-\overset{1}{N}H_2$ with CN on C3 $\xrightarrow{(1) \ C_5H_5N}$

[Pyridinone intermediate with $CH_2-O-C_2H_5$, CN, H_3C, N-H] $\xrightarrow{(2) \ HONO_2}$ [O_2N, $CH_2OC_2H_5$, CN, H_3C, N-H, =O] $\xrightarrow{(3) \ PCl_5, POCl_5}$ [O_2N, $CH_2OC_2H_5$, CN, H_3C, Cl]

$\xrightarrow{(4) \ 1. \ H_2, Pd \ \ 2. \ H_2, Pt}$ [H_2N, $CH_2OC_2H_5$, CH_2NH_2, H_3C, N] $\xrightarrow{(5) \ HONO}$ [HO, $CH_2OC_2H_5$, CH_2OH, H_3C, N]

$\xrightarrow{(6) \ H_2O-HCl}$ [HO, CH_2OH, CH_2OH, H_3C, N]

29.

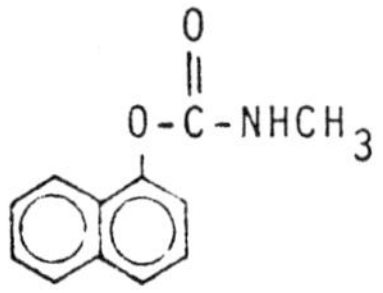

[Naphthyl-O-C(=O)-NHCH$_3$]

Sevin

$(CH_3)_2N$-[dimethyl-substituted aryl]-$O-\overset{O}{\underset{\|}{C}}-NHCH_3$

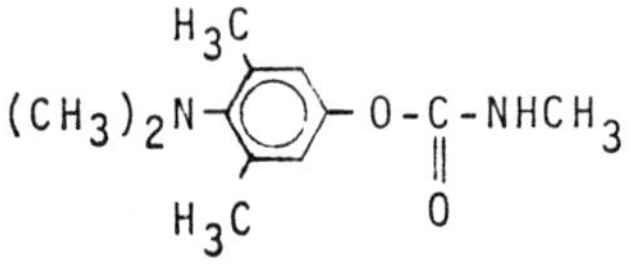

Zectran

30.

[Chlorophyll structure with rings A, B, C, D around Mg, substituents: CH=CH$_2$, CH$_3$, H$_3$C, C$_2$H$_5$, H$_3$C, CH$_3$, CH$_2$CH$_2$CO$_2$C$_{20}$H$_{39}$, CO$_2$CH$_3$]

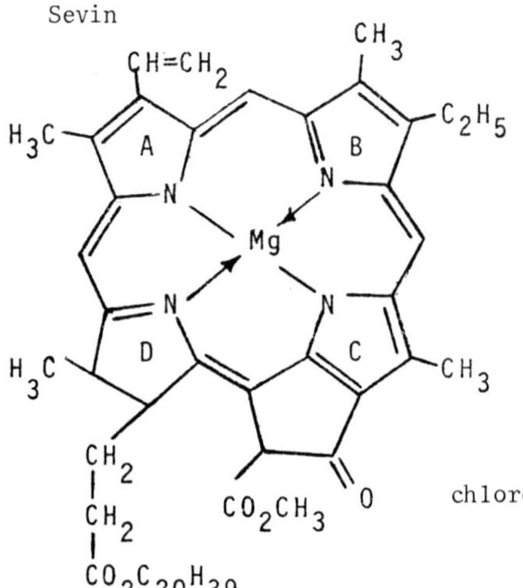

chlorophyll

chapter 19

DETERMINATION OF ORGANIC STRUCTURES FROM SPECTRAL PROPERTIES

1. Ultraviolet light is of higher energy than visible light. X-Rays are 10-10,000 times more energetic than even ultraviolet light, so that it is not surprising that x-rays can be injurious to life.

2. (1) 234 estimated, compared with 235 actual.   (2) 244 estimated, compared with 246 actual. Add 10 to 224 for each CH_3 group.

   1) $1/2.94$ = 3400, 1700, 810.   (2) $1/1815$ = 5.55, 3.33, 13.62.

   1) 1662 (conjugated C=O), 1638 (conjugated C=C), 1371 (CH_3—), 881, 808 (trisubstituted alkene). Only the bands which are useful in structural assignments are given. The molecular formula and spectrum suggest a cyclohexene ring with a methyl substituent. The low frequency of the carbonyl absorption indicates that the C=O group is conjugated with C=C. Because of the absorption for the trisubstituted alkene, it is clear that the methyl group must be attached to one of the carbon atoms joined by the double bond; which one cannot be determined. The spectrum is that of 3-methyl-2-cyclohexenone.   (2) 3560 (—OH), 1170 (tert OH); 3280 (—C≡C—H), 2290, 2175 (C≡C); 1380 (CH_3). The absorption bands are precise and definitive for the molecular formula. A tert hydroxyl group, a terminal alkyne group, and a methyl group could be joined in no other way to fit the formula 2-methyl-3-butyn-2-ol.
   (3) 3500, 3400, 1620, 1600 (primary amine); 758, 680 (monosubstituted benzene). Only aniline will fit the data.

5. (1) 1715 (from the table) - 10 (ring size correction) - 30 (conjugation, first C=C) - 15 (second C=C) = 1660 cm^{-1}.   (2) 1735 + 30 (ring size correction) - 30 (conjugation, first C=C) = 1735.

6. Since $\Delta E = \frac{hc}{\lambda}$, substitute in $\Delta E = \frac{\gamma hH}{2\pi}$. Thus, $\frac{hc}{\lambda} = \frac{\gamma hH}{2\pi}$ reduces to $\frac{c}{\lambda} = \frac{\gamma H}{2\pi}$. But $\frac{c}{\lambda} = \nu$, and, therefore, $\nu = \frac{\gamma H}{2\pi}$.

7. (1) $\overset{a}{C}H_3\overset{b}{C}H_2\overset{c}{C}H_2\overset{d}{O}H$, four signals.   (2) $\overset{a}{C}H_3\overset{b}{C}H_2-O-\overset{b}{C}H_2\overset{a}{C}H_3$

   (3) $\overset{c}{C_6H_5}\overset{a}{C}H_2\overset{b}{O}H$, three signals. Nonequivalent aromatic protons often give indistinguishable resonance peaks; therefore, an effort to differentiate these protons structurally is usually neglected.

(4) $\overset{b}{C}H_3-O-\overset{\overset{a}{C}H_3}{\underset{\underset{CH_3}{|}}{\overset{|}{C}}}-\overset{a}{C}H_3$, two signals (5) $\overset{a}{C}H_3\overset{b}{C}H\overset{c}{C}H_2Br$, three signals
$\underset{CH_3}{|}$

(6) b⟨a⟩b, two signals   (7) b{⟨$\overset{a}{OH}$⟩$\underset{a}{OH}$, two signals

(8) $\overset{a}{CH_3}\underset{Cl}{\diagdown}C=C\overset{\diagup H^b}{\underset{\diagdown H^c}{}}$, three signals. The nonequivalence of a proton may be established by substituting X for each H suspected of being the same to determine if the resulting formulas are different.

8. $CH_3\overset{O}{\overset{\|}{C}}-OC(CH_3)_3$: CH_3- 1.97; $(CH_3)_3C-$ 1.45. The alternative of

$(CH_3)_3C\overset{O}{\overset{\|}{C}}-OCH_3$ is ruled out because here CH_3- would show 3.6-5.1

(Table 19.5).

9. A lowering of deshielding of ring protons with consequent shift upfield is directly proportional to the number of electron-releasing methyl groups. (Circulation of π-electrons in aromatic rings and alkenes causes considerable deshielding of protons and chemical shifts well downfield (6.5-8.3 and 4.6-5.7, respectively).

10. $CH_3CH_2OH + e \longrightarrow CH_3CH_2OH^{\oplus \cdot} + 2e$; $CH_3CH_2OH^{\oplus \cdot} \longrightarrow CH_3\cdot + \overset{\oplus}{C}H_2OH$;
 46 (1) 31

(1) $\longrightarrow CH_3\overset{\oplus}{C}H_2 + \cdot OH$; (1) $\longrightarrow CH_3\overset{\oplus}{C}HOH + H\cdot$;
 29 45

(1) $\longrightarrow \overset{\oplus}{C}H_3 + \cdot CH_2OH$
 15

11. Structure (1) is a diene with four points of methyl or methylene substitution; hence from Table 19.2, we have an ultraviolet absorption maximum for butadiene of 217 nm. Each point of substitution causes a bathochromic shift of 5 nm; therefore, we would predict an absorption maximum at 217 + (4 x 5) = 237 nm. For structure (2), we would predict a maximim at a different wavelength. The α,β-

unsaturated ketone in Pb. 19.2 gives a value of 224 mµ with one point of substitution in the conjugated system. Since structure (2) also has one point of substitution, we would predict a similar value. Consequently a uv absorption spectrum should differentiate between the possibilities of (1) and (2) with values of about 237 and 224 nm depending on the structure. An infrared spectrum would add further evidence in that (2) would give a cyclic, conjugated carbonyl absorption band at about 1700 cm^{-1}.

12. The singlet at 10.95 indicates a carboxyl group, thus, $(C_3H_7O)COOH$. The other peaks center at about: (a) 1.27, (b) 3.66, (c) 4.13. The chemical shift at 1.27 suggests a methyl group; since it is a triplet, it would have a CH_2 (2 + 1) next to it. The CH_2 in turn is split into a quartet and because of the CH_3 (3 + 1), its other point of attachment must be to oxygen. Table 19.5 indicates that an ether O gives a chemical shift of 3.3-3.9; therefore, we can now write —O—CH_2CH_3. The structure could now be completed:

CH_3CH_2—O—$\overset{c}{CH_2}CO_2H$. The methylene group indicated as (c) would give

the peak furthest downfield because of the additional effect of the electron-withdrawing carboxyl group and, of course, there being no adjacent C atoms with protons, it would be a singlet.

13. (1) A CH_2 group flanked by 2 CH_2's, i.e., —$CH_2CH_2CH_2$—, would give a quintet (4 + 1). The molecular formula now simply involves insertion of an O:

```
  a┌────┐b
   │    │
  b└────┘O
```

The equivalent protons at b would give the observed triplet and would be further downfield because of the O.

(2) $\overset{a}{CH_3}\overset{b}{\underset{Br}{CH}}\overset{c}{CH_2}\overset{d}{CH_3}$

The methyl group (a) gives a doublet. The proton (b) gives a sextet downfield because of the Br. The CH_2 at (c) gives a quintet while the methyl (d) gives a triplet.

(3) $\overset{b}{CH_3}CO_2\overset{c}{CH_2}\overset{a}{CH_3}$

The triplet at 1.25 suggests a CH_3— group next to CH_2 (2 + 1, 0.9-2.2), while the CH_2 would give a quartet (3 + 1) and be next to an O (3.3-3.9). The singlet at 2.03 corresponds to the other CH_3— group.

(4) The molecular formula requires a benzene ring. A structure which would meet this requirement and three singlets would be:

c{⟨◯⟩—$\overset{b}{CH_2}\overset{a}{\underset{Cl}{C}}(CH_3)_2$

(5) $\overset{a}{CH_3}\overset{b}{CH_2}\overset{c}{CH_2}\overset{d}{CHO}$

The CH_3— group next to a CH_2 gives a triplet (2 + 1) at highest field. The neighboring CH_2 gives a sextet (5 + 1), while the other $\overset{c}{CH_2}$ gives a triplet downfield because of the electronegative aldehyde group. The latter gives a characteristic peak (d) at 9.74.

(6) The complex multiplet and molecular formula indicate a benzene ring:

$\underset{b}{CH_3O}-\underset{c}{\bigcirc}-\underset{a}{NH_2}$

An aromatic amino group (2.7-4.0) is indicated. The second downfield singlet is assigned to the CH_3O— group because of attachment to electronegative O. It is not possible to assign the orientation.

(7) The three singlets very neatly suggest benzyl alcohol as the structure:

$c\{\bigcirc\}-\overset{b}{CH_2}\overset{a}{OH}$

The alternative isomeric cresol structure is also a possibility. This might be questioned, however, in that the singlet at 4.58 is in a marginal position to be assigned to a phenolic hydroxyl group (4.5-8.0). The actual assignment for m-cresol shows the —OH at 5.67, CH_3— 2.25, and a complex multiplet at 6.48-7.23 for the benzene ring.

(8) $\overset{a}{CH_3}\overset{b}{CH_2}\overset{c}{CH}\overset{a}{CO_2H}$
              |
              Br

The splittings delineate the structure precisely along with the characteristic peak for the carboxyl proton.

14. (1) 1720 nonconjugated C=O; 1635 C=C; 1355 CH_3—; 985, 905 —CH=CH_2. 5-Hexen-2-one.   (2) 1795, 1743, two bands indicate carboxylic acid anhydride C=O; 758, 695 monosubstituted benzene. Benzoic anhydride.   (3) 3000 carboxylic acid O—H; 1725 carboxylic acid C=O; 1375 —CH_3. 2-Chloropropionic acid.

(4) 1685 conjugated C=O; 1355 —CH_3; 762, 695 monosubstituted benzene. Acetophenone (methyl phenyl ketone).   (5) 3225 —N—H; 1660 amide I absorption; 1585, 1550, 1440 amide II; 1365 CH_3—; 758, 690 monosubstituted benzene. Acetanilide or N-methylbenzamide.
(6) 1510, 1345 —NO_2; 790, 710, 678 monosubstituted benzene, nitrobenzene.

15. t-butyl oxalate